Tomorrow's Table

Organic Farming, Genetics, and the Future of Food

PAMELA C. RONALD
RAOUL W. ADAMCHAK

OXFORD
UNIVERSITY PRESS
2008

OXFORD
UNIVERSITY PRESS

Oxford University Press, Inc., publishes works that further
Oxford University's objective of excellence
in research, scholarship, and education.

Oxford New York
Auckland Cape Town Dar es Salaam Hong Kong Karachi
Kuala Lumpur Madrid Melbourne Mexico City Nairobi
New Delhi Shanghai Taipei Toronto

With offices in
Argentina Austria Brazil Chile Czech Republic France Greece
Guatemala Hungary Italy Japan Poland Portugal Singapore
South Korea Switzerland Thailand Turkey Ukraine Vietnam

Published by Oxford University Press, Inc.
198 Madison Avenue, New York, New York 10016

www.oup.com

Oxford is a registered trademark of Oxford University Press

Library of Congress Cataloging-in-Publication Data
Ronald, Pamela C.
Tomorrow's table: organic farming, genetics, and the future
of food / Pamela C. Ronald and Raoul W. Adamchak.
p. cm.
Includes bibliographical references and index.
ISBN 978-0-19-530175-5
1. Food—Biotechnology. 2. Genetically modified foods.
3. Organic farming. 4. Genetic engineering.
I. Adamchak, Raoul W., 1953– II. Title.
TP248.65.F66R66 2007
664—dc22   2007007071

1 3 5 7 9 8 6 4 2
Printed in the United States of America
on acid-free paper

Dedicated to Will Baker, 1935–2005
writer, farmer, and friend

FOREWORD

This book is a tale of two marriages. The first is that of Raoul and Pam, the authors, and is a tale of the passions of an organic farmer and a plant genetic scientist. The second is the potential marriage of two technologies—organic agriculture and genetic engineering.

Like all good marriages, both include shared values, lively tensions, and reinvigorating complementarities. Raoul and Pam share a strong sense of both the wonder of the natural world and how, if treated with respect and carefully managed, it can remain a source of inspiration and provision of our daily needs.

One of the greatest writers on agriculture was a Roman, Marcus Terentius Varro, of the first century B.C. In his classic book he described agriculture as "not only an art but an important and noble art."

It is, as well, a science. Not often do modern writers recall this fundamental truth. Raoul and Pam reflect it in their everyday lives. Raoul pursues the craft of organic farming, based on his experiences and those of farmers over the centuries, yet couples it with the modern science of ecology. For Pam, molecular and cellular science is paramount, yet she recognizes that all good plant breeders are also craftspeople in their day-to-day work.

The second marriage is more contentious: it tries to wed two entrenched camps where extreme views predominate. The marriage is long overdue. Several thousand years ago we humans had to give up hunting and gathering wild food sources. We began to domesticate and cultivate cereals and breed livestock. This process inescapably requires manipulation, which has grown increasingly complex and scientific.

Organic farming strives to maintain the centrality of natural processes—the value of organic matter as a source of nutrients and soil structure, and the role that natural enemies play in controlling pests, diseases, and weeds. Yet, as Raoul shows in this book, many of these processes have limitations in even a moderately intensive agricultural system. Pests, for example, may be very difficult to control. I know from my own work in Africa of the intractability of controlling the dreadful weed Striga or the pests and diseases of such crops as cowpeas and bananas using organic or conventional technologies.

What Pam and Raoul do is show that there is a role for genetic engineering in solving these particularly difficult-to-solve problems. Moreover, they show how technology can be applied in a way that strengthens organic farming performance and does not undermine its principles.

These are inspirational marriages.

Sir Gordon Conway, KCMG FRS, Professor of International Development, Centre for Environmental Policy at Imperial College, London, and past President of the Rockefeller Foundation

PREFACE

[An] emphasis on ecological processes and the complexities of household deci-
sion-making, may seem very distant from the molecular technology underlying
genetic engineering...; nevertheless not only are they both revolutionary in
their potential impact, they are interconnected.... The way forward lies in harnessing the power of modern technology, but harnessing it wisely in the interest
of the poor and hungry and with respect for the environment in which we live.
We need a shared vision, based, above all, on partnerships among scientists and
between scientists and the rural poor.

SIR GORDON CONWAY, *The Doubly Green Revolution,* 1997

By the year 2050, the number of people on Earth is expected to increase to 9.2 billion
from the current 6.7 billion (Population Division, 2007). What is the best way to
produce enough food to feed all these people? If we continue with current farming
practices, vast amounts of wilderness will be lost, millions of birds and billions of
insects will die, farm workers will be at increased risk for disease, and the public will
lose billions of dollars as a consequence of environmental degradation. Clearly, there
must be a better way to resolve the need for increased food production with the desire
to minimize its impact.

Some scientists and policy decision-makers have proposed that genetic engineering (GE), a modern form of crop modification (box P.1), will help create a new generation of plants that will dramatically reduce our dependence on pesticides, enhance
the health of our agricultural systems, and increase the nutritional content of food.
They believe GE will be a dramatic step forward that will allow agriculture to topple
decades of criticism about the dangerous overuse of pesticides and toxic herbicides,
leading us to a more ecological way of farming.

BOX P.1 **Genetic Engineering (GE)**

GE is not a farming method. It is a modern form of crop modification that differs from plant breeding in two basic ways:

1. Plant breeding allows gene transfer only between closely related species. With genetic engineering, genes from the same species or from *any* other species, even those from animals, can be introduced into a plant. Therefore genetic engineering creates a vast potential for crop alteration.
2. Plant breeding mixes large sets of genes of unknown function, whereas genetic engineering generally introduces only one to a few well-characterized genes at a time.

Or will it? While the public has generally accepted the application of GE for the production of new medicines, some consumers indicate grave unease over the consumption and production of GE food, viewing it as unnatural, potentially unsafe to eat and environmentally disruptive. Of these skeptics, the organic farming community has been particularly vocal in its criticism (box P.2). Some consumers believe that because organic farmers have learned how to produce healthy nutritious food, GE plants are not needed.

BOX P.2 **Conventional and Organic Farming**

Conventional agriculture is a catch-all term used to describe diverse farming methods. At one end of the continuum are farmers who use synthetic pesticides and fertilizers to maximize short-term yields. At the other end are growers who use chemicals sparingly and embrace the goals of ecological farming. Increasingly, many conventional farmers, particularly in the United States, are growing GE crops.

Organic farming is an ecologically-based farming method that avoids or largely excludes the use of synthetic fertilizers and pesticides. As much as possible, organic farmers rely on crop rotation, cover crops, compost, and mechanical cultivation to maintain soil productivity and fertility, to supply plant nutrients, and to control weeds, insects, and other pests. The United States Department of Agriculture (USDA) National Organic Program standards established in 2000 prohibit the use of GE seed or other GE inputs. Currently, organic farming is practiced by less than 2% of U.S. farmers.

Over the last ten years of marriage, we, Raoul Adamchak (an organic farmer) and Pamela Ronald (a geneticist), have discussed these issues with each other and with others. We both work at the University of California at Davis, a world-class research institution that is located amid some of the world's richest soils in the fertile Central Valley. An unusually high percentage of the people who live in the small town of Davis studies or cultivates plants. Here, organic growers and geneticists routinely mingle together in the same social circles. Many of our friends, family, and colleagues see GE and organic farming as representing polar opposites of the agricultural industry, and they often ask us how GE will affect the environment and our food. On the other hand, some of our scientific colleagues have asked us to explain why many people in the organic farming community oppose the genetic engineering of crops. This book is the result of our investigations and our response to these questions.

Written as part memoir, part instruction, and part contemplation, this book roughly chronicles one year in our life. Our intention is to give readers a better understanding of what geneticists and organic farmers actually do and also to help readers distinguish between fact and fiction in the debate about crop genetic engineering. Readers who wish to know more about the science behind the passionate arguments surrounding genetic engineering and organic agriculture can find it in this book.

One of the major themes of this book is that the judicious incorporation of two important strands of agriculture—genetic engineering and organic farming—is key to helping feed the growing population in an ecologically balanced manner. We are not suggesting that organic farming and GE alone will provide all the changes needed in agriculture. Other farming systems and technological changes, as well as modified government policies, undoubtedly are also needed. Yet it is hard to avoid the sense that organic farming and genetic engineering each will play an increasingly important role, and that they somehow have been pitted unnecessarily against each other. Our ambition in this book, therefore, is not to be comprehensive, but to identify roles for both GE and organic farming in the future of food production.

Another theme of this book is that the broader goals of ecologically responsible farming, and the adherence to those ideals, are more important than the methods used to develop new plant varieties. To this end, we have generated a list of key criteria to help guide policy decisions about the use of GE in food and farming (box P.3). Throughout this book, we evaluate the usefulness of a particular crop variety or farming technique using these criteria. By looking beyond the ideologies and ahead to a shared vision, we hope to better achieve these goals.

BOX P.3 **Criteria for the Use of Organic Farming and Genetic Engineering in Agriculture**

We advocate the use of a technology or farming practice if it serves to:

- Produce abundant, safe, and nutritious* food
- Reduce harmful environmental inputs

(continued)

- Provide healthful conditions for farm workers
- Protect the genetic make-up of native species
- Enhance crop genetic diversity
- Foster soil fertility
- Improve the lives of the poor and malnourished
- Maintain the economic viability of farmers and rural communities

* As defined by the United States Department of Agriculture Food and Nutrition Service

Loosely organized by season, each group of chapters addresses a different issue related to the role of GE and organic farming in food production. For example, chapter 1, written by Pam, is a case study showing how plant geneticists are working with breeders to address agricultural problems faced by farmers in less developed countries. Chapters 2 and 3, written by Raoul, provide a farmer's-eye view of the philosophy and practice of organic farming and how it differs from conventional agriculture. Chapters 4, 5, and 6, written by Pam, describe the tools and processes of genetic engineering, examine consumers' concerns and review the scientific process. In chapters 7, 8, and 9, Pam discusses potential health and environmental risks and benefits of GE crops. In chapters 10 and 11, we discuss the role that private companies and patents play in the development of new seed varieties. The last chapter describes a typical California spring dinner that we prepare for our family. Some of the food is genetically engineered and some is grown organically. We explain why we make the choices we do. Because our book is essentially about food, we include some of our favorite recipes.

We wrote this book for consumers, farmers, and policy decision makers who want to make food choices and policy that will support ecologically responsible farming practices. It is also for consumers who want accurate information about genetically engineered crops and their potential impacts on human health and the environment. Our book is for those who wish to know more about the food they eat, besides just how to prepare it. It is for every shopper who has at one time or another perused the aisles of the local supermarket wondering what labels such as "organic" or "GE-free" really mean for the health of their families and for the future of the planet.

Acknowledgments

Many people made vital contributions to the research and writing of this book. They include Amie Diller, Cindy Toy, Matt Ariel, Kim Barnes, our editor Peter Prescott and Pam's assistant Rebecca McSorley. These friends patiently read numerous drafts, provided critical comments that improved the book and gave us the encouragement needed to start, as well as complete, the project.

We are also grateful for the support of many other friends, family, and colleagues who lent their valuable time to this project. Their questions, unique expertise, and perspectives helped us to be accurate, thorough, and balanced. They include Frances Andrews, Julia Bailey-Serres, Will Baker, Diane Barrett, Laura Bartley, Alan Bennett, Kent Bradford, Mateo Burtsch, Patrick Byrne, Patrick Canlas, Gordon Conway, Doug Cook, Joe DiTomaso, Aliki Dragona, Bryce Falk, Sean Feder, Sally Fox, Gina Fregosi, Paul Gepts, Val Giddings, Leland Glenna, Claire Gelfman, Anne Harper, Susan Harrison, John Hill, Paul Holmes, Junda Jiang, Gurdev Khush, Peggy Lemaux, Rachel Long, Jessica Lundberg, Pamela Martineau, Claire Mazow Gelfman, Cora Monce, Jane Miller, Elise Pendall, Ginny Powers, Stan Robinson, Patricia Ronald, Peter Ronald, Rick Ronald, Robert Ronald, Ruth Santer, Barbara Schaal, Kim Shauman, Alycia Somers, Sharon Strauss, Bruce Tabashnik, Beth Tsauzik, Allen Van Deynze, Alison Van Ennenaman, Mark Van Horn, Andrew Waterhouse, Victoria Whitworth, Valerie Williamson, Kenong Xu, Xia Xu, and past and present members of the Ronald Laboratory Group and the U.C. Davis Student Farm.

Finally, thanks to Cliff and Audrey for their patience as we typed away ("If the book is done you won't have to spend so much time at the computer, right?") and to our parents Julie and Bill Adamchak and Trish and Robert Ronald. Without their love and generosity, this book would not have been possible.

NOTE Nearly all the people and events described in this book are real. Some names have been changed, a number of events have been merged, and a few experimental methodologies simplified.

Contents

About the Authors

Raoul Adamchak has grown organic crops for twenty years, part of the time as a partner in Full Belly Farm, a private 150-acre organic vegetable farm that provided weekly produce boxes to over five hundred subscribers. Raoul has sold produce at three high-volume farmers' markets, and to wholesalers and retailers in the San Francisco Bay Area and Sacramento. He has also spent many hours discussing organic certification issues as a member and president of California Certified Organic Farmers' (CCOF) and Board of Directors and inspected over one hundred organic farms for CCOF. He received a bachelor's degree in economics from Clark University and also received a master of science degree in International Agricultural Development from the University of California, Davis, where he also studied entomology. He now works at the University of California, Davis Student Farm, where he teaches organic production practices and manages a five-acre market garden.

Pam Ronald is Professor of Plant Pathology and Chair of the Plant Genomics Program at the University of California, Davis, where she studies the role that genes play in a plant's response to its environment. Much of her work has focused on rice, a staple for 50% of the world's people. Her laboratory has genetically engineered rice for resistance to diseases and flooding, both of which are serious problems of rice crops in Asia and Africa. She received a bachelor's degree in biology from Reed College, a master of science degree in plant physiology from Uppsala University in Sweden, a master of science in biology from Stanford University, and a Ph.D. degree from the University of California, Berkeley. Her work has been published in *Science, Nature,* and other scientific periodicals and has also been featured in newspapers including the *New York Times,* the *Wall Street Journal,* and *Le Monde.* Ronald is a 1984 Fulbright Fellow, a 1999 Guggenheim Fellow, and an elected Fellow of the American Association for the Advancement of Science.

Introduction

One

Cultivating Rice in Nihe, China, and Davis, California

Oh sacred padi, You the opulent, you the distinguished, Our padi of highest rank;
Oh sacred padi, Here I am planting you; Keep watch o'er your children, Keep watch
o'er your people, Over the little ones, over the young ones, Oh do not be laggard, do
not be lazy, Lest there be sickness, lest there be ailing; You must visit your people,
visit your children. You who have been treated by Pulang Gana; Oh do not neglect
to give succour, Oh do not tire, do not fail in your duty.

Words of prayer addressed to the rice (*padi*) spirits by an Iban farmer in Borneo
circa 1950. The god of fertility (Pulang Gana) is invoked because the community's rice crop was failing to thrive.

As quoted in D. Freeman, *Report on the Iban*, 1970;
and R. W. Hamilton, *The Art of Rice,*
Spirit and Sustenance in Asia, 2003

The flooded field was drained last week. Today, I trudge through the mud, feeling
the cool, wet clay pass between my naked toes; my straw hat shades my face from the
hot sun of the typical Davis summer day. The dark mud pulls strongly at my heel,
releasing my foot with a loud sucking noise just in time for the next step. Because my
shoes have been lost before in these fields, I have left them behind. A great blue heron
flies nearby; her squawking frightens a flock of small white egrets, which lifts up in a
panic. It is rice that draws me here today—the crop that has fed more people over a
longer period of time than any other (Huke and Huke 1990).

I don't get out to the rice field much anymore, but today is different. This morning
Kenong and his wife Xia, who work as researchers in my lab, plan to show me the
result of an important experiment. I am impatient to see it. For over ten years we have
been trying to identify a rice gene that is critical for keeping rice plants alive under
flooded conditions. Although scientific exploration is a slow process, we draw a kind
of broad satisfaction from it, for with each new discovery, each new bit of scientific
information, we are a step closer to understanding a piece of the natural world.

The experimental field is on the UC Davis campus, situated in the 450-mile-long Central Valley of California. Like most rice farms in California, the ground is leveled with laser-controlled precision to ensure that the crop is submerged in water to exactly the desired depth to discourage weeds. Water level in the fields is controlled through inlets and outlets as if each paddy were a shallow bathtub being filled or drained by the turn of a faucet or the pull of a plug. Farmers vary the level of water to control weeds. Weeds that are covered by deep water are cut off from the supply of air and light needed to maintain respiration and photosynthesis and will die. This approach, however, is not 100% effective. I see small-flowered umbrella plants, white-flowered ducksalad and barnyard grass, all weeds, thriving quite well in parts of the paddy where the water is not very deep. Because of these recalcitrant weeds, farmers often also use other strategies such as chemical herbicides to kill the weeds. This secondary approach cannot be used by organic growers, who rely heavily on the use of water to control weeds because they are prohibited from using herbicides. The problem is that if the young rice seedlings are completely submerged for more than a week, they will die. Thus, the organic farmer walks a fine line between killing the weeds and killing the rice.

The pungent scents from the surrounding dry grass and standing water are characteristic of the Central Valley in summer, bringing to mind tramps with my brothers in the nearby coastal range where, as children with ample leisure time, we would use cardboard sleds to slide down grass hills and plastic buckets to collect polliwogs from an abandoned well.

As we walk through the mud, Kenong tells me about a very different childhood. He was born in 1963 in Nihe, a small village nestled in a valley east of the Daibe Mountains in the Chinese province of Anhui (figure 1.1). Much of Anhui is a large flood plain of the Yangtze River, which provides fertile ground for growing rice. "I lived with my parents, brother and two sisters in a small brick house. My family had fifteen hens, five roosters, fifteen ducks, two pigs, and ten geese. We did not have enough grain to feed the animals, so they would scrounge what was available: the pigs eating table scraps, the poultry pecking for bugs and greens, and the ducks catching the occasional fish in the nearby river. Once the animals were big enough, we would kill and eat them, adding variety to my family's three times daily diet of rice, rice, and rice. For play, my brothers and sisters and I caught fish with our hands from the local river, which flows into Caohu Lake and then eventually into the Yangtze fifty to sixty miles away."

The thirty families in the village worked collectively to care for the rice crops. Every day before school and in the summer, Kenong would help. Early in the spring, the women would transplant the seedlings into the wet paddy, bending close to the ground to be sure that each seedling was unbroken and its roots buried deeply in the mud. Throughout the growing season the fields were laboriously hand-weeded. It was backbreaking work, but essential because the weeds competed for light, water, and

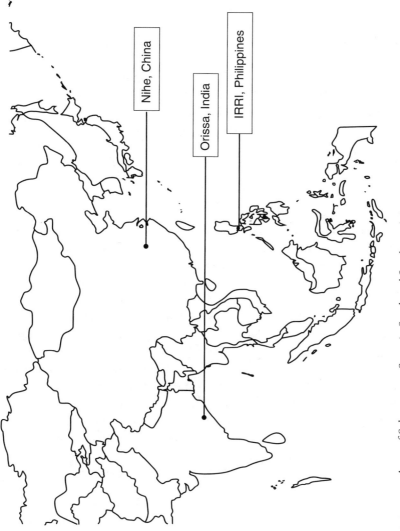

Nihe, China

Orissa, India

IRRI, Philippines

FIGURE 1.1 Areas of Submergence Stress in South and Southeast Asia.

nutrients. In the fall, the men would heave great bundles of rice onto their shoulders and walk to town to sell them. Trucks and tractors were rare sights.

During Mao Tse-tung's "Great Leap Forward," a mass campaign to collectivize agriculture and speed industrial growth from 1958 to 1962, tens of millions of Chinese died of starvation. Yet, because they farmed on some of the most productive land in China, the Nihe villagers usually had enough to eat. Kenong recalls the one exception.

"I remember a terrible time in my village. When I was six or seven years old, a flash flood rolled down from the Daibe Mountains submerging the freshly transplanted field. The entire crop was destroyed. Because my father was a salaried government employee, my family had enough to eat. Most of our neighbors, however, were not so fortunate. When the government and neighbors could not provide enough food, many of them starved."

The Nihe villagers were not alone—each year, an estimated 15 million hectares of rice lands (a region half the size of Italy) in South and Southeast Asia are inundated by flash floods. In Bangladesh during the monsoon, roads are so wet that they become waterways for homemade sailboats rigged with cloth, jute, and bamboo (Catling 1992). Such lands are home to an estimated 140 million people of whom 70 million are living on less than $1 a day, the highest concentration of poor people in the world. Here, losses of rice production can be over $1 billion per year (Herdt 1991; Dey and Upadhyaya 1996). This number, however, does not capture the human suffering caused by the catastrophic crop losses.

Although rice is the only cereal that can withstand some flooding, most rice varieties will die if submerged for too long. There are a few rare exceptions, and these are of great interest to rice breeders (scientists who mix and match genes between rice varieties to generate an improved strain). One of these is the traditional Indian rice variety, FR13A. This rice plant has an unusual and agronomically important trait—the seedlings are able to withstand fourteen days of submergence. It is, however, low yielding and no longer widely grown. FR13A originated in the state of Orissa, in eastern India, bordered on the east by the Bay of Bengal. Hindu temples dating to the thirteenth century are scattered through the area. Today most of the people there still speak the ancient dialect of Oriya, and the majority are still rice farmers. For over fifty years, breeders tried to use FR13A as a parent plant to introduce the submergence tolerance trait into high yielding, tastier varieties favored by rice farmers in other parts of Asia. Frustratingly, the resulting new varieties were of poor quality. The main reason for this breeding failure was that, because they were not really sure which genes were needed or where in the genome they were located, the breeders accidentally introduced other genes that reduced the overall quality of the rice (box 1.1).

BOX 1.1 **Genes and Genomes**

What is a gene? A gene contains hereditary information encoded in the form of deoxyribonucleic acid (DNA) or ribonucleic acid (RNA). DNA is a long, stringy substance composed of four different nitrogen-rich compounds (called adenine, thymine, guanine and cytosine), sugars, and phosphorus. Each gene codes for a single protein or a family of related proteins. Genes determine most aspects of anatomy and physiology by controlling the production of these proteins. In higher organisms, such as rice and humans, genes are located at a specific position on a chromosome in a compartment of the cell called the nucleus.

What is a genome? How many genes are in the rice genome? The genome consists of all of the genetic information or hereditary material possessed by an organism. The finished sequence of the rice genome was completed in December 2004 by the International Rice Genome Sequencing Project, a consortium of publicly funded laboratories. Sophisticated computer programs were used to scan the sequence to identify candidate genes. Using this method, approximately 42,000 genes were identified. Many of these genes appear to be unique to cereals. One of the goals of modern rice genetic research is to determine the function of every one of these genes and then to use that information to advance our understanding of basic biological processes, as well as to enhance crop productivity.

Kenong, Xia, and I stop to look around the field. The even stands of tall green grass-like leaves wave slightly in the wind, as if they were friends welcoming us to a feast. For farmers, such a sight is welcome, because it means many plants that will yield abundant grain.

"How did you get interested in rice breeding?" I ask Xia and Kenong.

"We both wanted to grow rice," Xia answers. "And we both wanted to go back to school."

Kenong explains, "During the Cultural Revolution, Mao Tse-tung closed schools, and we did not have a chance to study. When the schools opened again, agriculture was a very popular subject, so many of us enrolled at Anhui Agricultural University. That is where we met."

Xia says, "I came from the north, from the county government seat near Mount Huangshan. This is a very famous place. It is known for its many cliffs, pine trees, and hot springs. We then moved to Wuhan, China, where Kenong received his M.S. in agronomy. That is where our first son was born, and also where we learned about Professor Dave Mackill who was working on rice submergence tolerance."

At that time, Dave was a genetics professor and colleague of mine here at UC Davis. He was well known for his efforts to apply modern genetic techniques to rice breeding. Over the years, Dave has patiently answered my questions about rice breeding and farming, explaining the problems encountered by Californian and Asian rice farmers. Before coming to Davis, Dave worked at the International Crop Research Institute for the Semi-Arid Tropics, in Patancheru, India, and then spent ten years at the International Rice Research Institute (IRRI) in the Philippines. As a researcher, Dave's main goal has been to develop new rice varieties that will help poor farmers combat agricultural problems such as submergence, drought, and disease. He thought that if he could isolate the submergence tolerance gene from FR13A, breeders would be able to precisely introduce it into rice varieties that were killed by submergence, helping farmers in Asia and possibly in California as well.

Dave's vision and leadership brought Kenong, Xia , and me together. In 1996, after Kenong and Xia joined his lab, Dave asked if I would use my expertise in rice genetics to help them identify the submergence tolerance gene from FR13A. Within a couple of years, Kenong and Xia were able to locate the submergence tolerance trait to a very small region of one of the rice chromosomes (Xu et al. 2000). Computer programs allowed us to predict the function of the genes in this region, and one was of particular interest. Based on what we knew about this kind of gene from other plants, we hypothesized that it might act as a master switch to regulate complex functions of the plant. It was as if Kenong and Xia had been able to unravel a ball, woven from 42,000 silken threads all of a slightly different hue, and to pull out one thread, interlaced but distinct from the others. Unlike weavers, geneticists cannot determine if the thread they hold is the one they want simply by looking at it; instead they need to test it by weaving it into another pattern—in this case another rice plant that normally cannot survive floods (for details, see figure 4.4). So that is what we did. We genetically engineered (GE) this single thread, carrying the submergence tolerance trait, into a rice variety that normally would die in a flood. We wanted to know if incorporation of this one gene would allow the plant to survive.

Kenong and I follow Xia through the paddy. Although she is much shorter than us, we must walk quickly to keep pace. It is she who transplanted the young GE seedlings several weeks earlier.

We submerged the seedlings in this paddy for over two weeks, and then drained it ten days ago. Only a few rice plants have survived the flood and the ones I see as we walk through the field are weak, spindly and very pale. This pale and flaccid appearance is typical of plants that have drowned, lacking the air and sunlight needed to function. It is unlikely that this group of plants will survive much longer.

In the distance I see a few rows of resilient rice seedlings rising up out of the black mud. Xia tells me that these are the plants that carry the genetic information from the submergence tolerant Indian variety. If we have identified and introduced the correct gene, the plants will have survived the extended time underwater and recovered. I hurry over, bend, and gently touch the bright green leaves of the first plant. My eyes quickly travel down the row. They are all alive. It is as though the rice plants had been able to hold their breath until the water was gone. Like magic. I look up into the faces of Xia and Kenong, their smiles reflecting a sense of joy and a feeling that all will be right in the world. We excitedly head back to the lab to tell the others, the mud pulling on our bare heels.

I realize that I would not have known Kenong and Xia but for this communal weaving of an ancient thread of DNA into the modern rice variety. Our work represents the latest genetic change in the rice plant, which was first cultivated along the Yangtze River 6000 years ago (Huke and Huke 1990). Since that time, hundreds of thousands of rice varieties have been developed. It is likely that FR13A was selected by Orissan farmers because it could survive the floods particular to that area. It was then handed down from one generation to the next, prized then, as now, for its submergence tolerance. Our latest work tells us that the submergence tolerance trait is found not only in the Orissan variety, but also in two traditional varieties from Sri Lanka. It appears that ancestors of the Sinhalese, who originated from Orissa and migrated to the island twenty-five hundred years ago, transported these precious rice grains over thousands of kilometers.

Perhaps as geneticists, we are acting as humans have always done: learning the secrets of the sacred and ancient and passing that knowledge to others, who will then use that information in a new and unexpected way. The submergence tolerance gene has now returned to southern Asia in another new form. With the use of modern genetic techniques (box 1.2), Dave and coworkers have introduced this gene into rice varieties that are adapted to habitats in South and Southeast Asia (Xu et al. 2006). The new plants can withstand fourteen days of submergence, and they yield and taste the same as their parent (figure 1.2). Production trials in Laos, Bangladesh, and India are now well underway. These newly modified rice plants are expected to help alleviate the suffering of poor farm families in Asia. Because our team has also created California rice varieties carrying the submergence tolerance trait, we may be able to help our local organic rice growers and other farmers fight weeds without herbicides. If so, then once again, humans have muddied the lines that separate the traditional and the modern; the farmers of the past with those of today; the organic and the genetically engineered.

What we know for sure is that agriculture will continue to change, and that we are changing along with it—our lives touched, slightly altered, our perspectives renewed by our work and those with whom we work and live.

FIGURE 1.2 Submergence Tolerant Rice.

Fourteen-day-old seedlings were submerged for fourteen days and photographed fourteen days after desubmergence. On the left is the submergence intolerant variety, Swarna, which is well-adapted to farms in South and Southeastern Asia. On the far right is a rice line carrying the submergence tolerance trait. The two varieties in the middle are genetically similar to Swarna except for the addition of the submergence tolerance gene *Sub1*, introduced using marker-assisted breeding (Xu et al. 2006).

BOX 1.2 **Genetic Modification of Rice Using Marker-Assisted Breeding**

With marker-assisted breeding, researchers first identify the genetic "fingerprint" of the genes that they would like to move from one variety to another. Then, using traditional plant breeding practices, the two varieties are cross-pollinated, producing many offspring. Next, the breeder identifies which of the offspring appears to have the desirable genetic fingerprint, and not the undesirable ones. The process is then repeated. Without the genetic fingerprint and confirmation of useful genes by genetic engineering, selecting for complex traits is extremely difficult. With marker-assisted breeding researchers can look directly at the DNA of the offspring to determine exactly what has been inherited. Marker-assisted breeding is more widely accepted than GE because it is considered by most countries to be an extension of conventional plant breeding although it relies on modern genetic techniques.

For the FR13A project, marker-assisted breeding was very helpful because researchers could screen for the best varieties in the lab, instead of having to flood a rice field four feet deep to see if the offspring tolerate submergence, thus saving a lot of time and labor.

The Farm

Two

Why Organic Agriculture?

It is early July, the first week of our class in organic production practices and even at 9 A.M. the morning sizzles. We have eight new students, and this twenty-acre certified organic farm on the UC Davis campus is our training ground. At the end of the class, those students who manage to persevere through the hot Davis summer will have learned how organic production differs from conventional practices and how to farm organically.

As an introduction, I take the class for a tour of the market garden. We walk between the rows of melons: Sivan, Stutz Supreme, and Crimson Sweet. The students might see diversity and color, I see more work projects. We move to the next rows. There, early tomatoes have been pruned and staked. The first level of string runs from stake to stake to support the growing plants. Over the summer the students will tie the next four levels so that the ripening fruits do not sink to the ground. The reward will be red, yellow, and purple fruit appropriate for a multi-colored pizza. The beds at the far end of the field look like the cloth-covered furniture at my Aunt Olga's house. This is basil, ready to harvest, but still covered with a floating row cover to keep the flea beetles from eating the leaves. I explain that the fruits and vegetables grown here will be sold to UCD staff, faculty, and students, as well as to the student-run coffee house on campus. Eventually we end up in the shade of a massive fig tree, where I intend to begin my lecture.

But first I invite everyone to taste the Black Mission figs, succulent and sweet. Their soft, almost gooey insides look like hundreds of short, reddish tentacles. Although considered a fruit, the fig is actually a flower inverted into itself. When people from the Middle East visit the farm, I notice that they peel the fig before eating it. Today, the class follows my example and eats them whole—skin and all. Like bees calmed by honey, a class gorged on figs is soon ready to listen.

"Organic farming came about as a response to the environmental and health problems associated with overuse of chemicals on conventional farms," I explain as I notice that the students are cooling off, getting comfortable and beginning to pay attention. "It is best described as 'better farming through biology' because it is based on using living organisms rather than synthetic chemicals."

"Conventional farming, in contrast, could be described as 'better farming through chemistry,' because many conventional farmers use synthetic pesticides and fertilizers that became readily available in the 1940s (box 2.1). The goal of conventional farming is high yields and inexpensive food. The goal of organic farming is health: health of the soil, the crop, the farmer, the environment, and the consumer."

Katie, one of the students who must have been looking at some of the weedier parts of the field while I was talking, raises her hand and asks, "That sounds good, but how do you control weeds without herbicides?"

Weeds can be a sore point on organic farms. Crop rotation, mechanical cultivation, pre-irrigation, and the removal of weeds before they go to seed, are all weed-reduction strategies used on organic farms. They work well when there is enough labor to stay on top of everything that needs to be done. When there is not enough labor, weeds can be a problem. Another strategy, which is more technological than biological, is soil solarization. In the 1980s Clyde Elmore, a UC Davis professor, demonstrated that covering moist soil with a thin, clear plastic for six weeks in the heat of the summer kills nearly all the weed seed in the top inch and a half of soil (Elmore 1997).

I use soil solarization for crops like carrots that are the most sensitive to weed competition, and the most labor intensive to weed. Since experiential learning is the teaching style of the class, (the students learn concepts by putting their bodies to work), to answer Katie's question I ask the class to follow me back into the heat so we can get to work using this method of controlling weeds without herbicides.

BOX 2.1 **Types of Pesticides**

Herbicides: kill weeds
Insecticides: kill insects
Fungicides: kill fungi
Bactericides: kill bacteria
Nematocides: kill nematodes

Weeds, insects, bacteria, nematodes, and fungi are major pests of crops and are estimated to reduce the global yield by 40% annually.

Our exercise today is to cover sixteen beds of freshly tilled and moist soil in clear plastic. Unfortunately, I left purchasing the plastic to the last minute and was only able to find rolls thirteen feet wide. Not the perfect size, because the beds are six feet wide and the tractor-mounted sled is meant for six-foot-wide rolls. With the too-wide rolls in hand, I realize I will need to cut the rolls in half. The plastic is, however, folded diabolically—overlapping itself in such a way that makes it very time-consuming to unfold and cut, which is not ideal on a hot day. I then try to figure out a way to feed the plastic we've cut into the sled. I am winging it, and it is not working. The bright sun is dazzling as it reflects off the unrolled plastic sheeting. I am sweating, everyone is sweating, and we are quickly losing patience.

One of my students, Sang Min Lee, coolly watches my efforts. Most of the time, when you are the instructor in a class, the students know much less than you and have so little experience that they just do what you say, even if it's a little wild and crazy. Sang Min Lee is an exception. Unlike the typical young and idealistic student, who views organic farming as a necessary ecological alternative to conventional farming, Sang Min is a grower of conventional crops and is mostly interested in the profit potential of organic production. He is in his sixties, which is hard to believe because he looks 45 and works like a 35-year-old. His company, Daikon Farms, grows or contracts to grow thousands of acres of Asian melons and daikon radishes. He is a farmer, businessman, poet, teacher, and ex-ship captain.

After giving me a little time to dig myself out of the abyss, Sang Min says, "You don't have to cut in half. Use whole piece. Don't use machine." And I can't argue with that. The students and I roll out another section and leave it wide. It covers two beds. We scoop shovels full of soil every five feet along the outside edges and throw a few clods into the middle furrow to seal in the moisture. It takes about ten minutes. We do another two beds, ten more minutes. In an hour we cover ten beds. It's time for the class to move on to the next lesson, and I am elated by our progress.

I next take the students into the organically managed greenhouse to show them how to plant seeds. The air smells faintly of the fish emulsion that was recently used to fertilize the seedlings; I have to shout over the rush of air created by the evaporative coolers. "There are advantages of sowing seeds into trays instead of directly into the soil. While the seedlings are growing in the greenhouse, we can irrigate the fields to germinate weeds. When the weeds are small, we shallowly till to kill the weeds. When the crop seedlings are large enough, we can transplant them into the soil, and they will have a big head start against the weeds."

While there are machines that can place seeds in all 128 cells of the tray simultaneously, I have the students sow the seeds by hand. It is a little slower, but is a contemplative process that is conducive to conversation. As the class sows the pepper, tomato, basil, and eggplant seeds, the questions emerge out of the silence. Sang Min, who has happily been using synthetic fertilizers for many years, asks, "What is wrong with chemical fertilizers? And, what is the organic alternative?"

I start from the beginning with my answer explaining that plants need nitrogen, phosphorus, and potassium, as well as calcium, sulfur, magnesium, and a number of micronutrients to grow and thrive. Nitrogen is an essential component of both DNA and protein, which are necessary for virtually all plant processes. Low amounts of nitrogen in the soil result in stunted plants. Phosphorus has many roles; it is needed for the conversion of light energy to chemical energy (ATP) during photosynthesis, and also plays an important role in the way that the plant transmits signals throughout the cell. For instance, when a plant senses a potential threat, it uses phosphorous to signal the cell that defense mechanisms should be deployed. Potassium activates enzymes that control plant functions and is important in controlling stomata (pore-like openings) in leaves. The model in conventional agriculture has been to feed the plant synthetic fertilizers. In contrast, the organic model seeks to provide the soil with nutrients that will be needed by the plant, and let the microorganisms in the soil "mineralize" those organic materials into a form that can be taken up by the plants.

I go on to explain how conventional farmers use N, P, K fertilizers (so called for the nitrogen, phosphorus, and potassium they contain) that have been synthetically produced using fossil fuels. It takes the energy equivalent of thirty gallons of gasoline to produce the synthetic N, P, K needed to grow an acre of corn (Shapouri et al. 2004). The use of synthetic fertilizers has increased forty-fold in the last fifty years, and is still rising. Increasing energy costs will severely impact conventional agriculture. Recently, on CBS news, a conventional farmer in Georgia was complaining that fertilizer costs had gone up 48% in the last three years. This had added $54,000 to his costs in 2006. (CBS, May 6, 2006).

Organic farms use less energy to generate nitrogen fertilizer because much of it comes from cover crops, instead of petroleum-based sources. Cover crops are plants that are grown to be turned back into the soil for nutrients and organic matter. The two cover crops we use most often, vetch and bell beans, are planted in the fall and tilled under in the spring. These plants belong to the legume family and have a symbiotic relationship with bacteria that lives in the soil. Remarkably, this partnership draws nitrogen from the air and converts it to a form of nitrogen that plants can use. Legume cover crops can add as much as 150 pounds of nitrogen per acre, which is enough to support the growth of a variety of summer vegetable crops. We can also plant summer-grown legumes, like cow peas, that provide the nitrogen needed for winter crops. In on-farm cover crop trials done near Davis, a winter crop of vetch used between 46% and 67% less energy than fields fertilized with chemical fertilizers. (Klein 1989).

David Pimentel, a Cornell University professor of ecology and agriculture analyzed a 22-year organic versus conventional farming trial done at the Rodale Institute in Pennsylvania. He concluded that organic farming produced the same yields of corn and soybeans as conventional farming, but used 30 percent less energy (Pimentel 2005). Scientists at the Research Institute of Organic Agriculture in Frick, Switzerland, found that energy inputs to the organic system were about half of those to the conventional system and used 97% fewer pesticides, although yields were considerably lower (Maeder et al. 2002).

But reduced energy use is only part of the story. Synthetic N, P, K are very soluble in water. This makes it easy for plants to use them, but synthetic fertilizer that is not taken up by the plants readily runs off or leaches out of the field into streams or into groundwater. The consequences have been serious. Excess nutrients in lakes and rivers cause algae to multiply and use up all the oxygen, resulting in death of fish and shellfish. Each summer, a 6,000-square-mile dead zone forms at the mouth of Mississippi River due to high levels of nitrogen and phosphorus drained from thousands of acres of farmland upstream (Texas A&M University News, April 6, 2005). On organic farms there is 4.4 to 5.6 times less nitrate leaching than on conventionally farmed fields (Kramer et al. 2006). This is because a large percentage of the nitrogen is bound within organic molecules, and it only becomes available to plants as the organic molecules are broken down into soluble ions by soil microbes and worms.

"Cover crops good, but no money," Sang Min says suddenly. All the students turn to look at me, waiting for an answer.

Startled, I glance over at Sang Min. "It is true that when cover crops are in the ground, money-making food crops can't be grown in the same space. However, cover crops also have other critical roles. They help suppress weeds, deter the build-up of insect pests (if they are not a host of the same pests that attack the crop), and add organic matter to the soil. This added organic matter enhances microbial activity and builds soil structure. Cover crops are making money for growers, but in indirect ways."

Rotating to a cover crop also helps reduce insects and nematode pests, weeds, and plant disease. When a grower doesn't rotate, he is likely to be faced with one or more of these problems. Conventional growers can use pesticides to control these pests rather than rotate. So there is a choice between crop yield (while cover crops are planted) and pesticide use (that can maintain yield but may have toxic effects on the environment), at least in the short run. In the longer run, pests become resistant to pesticides and farmers may be forced to "rotate" because the crop can no longer be grown in that area.

Another student, Jim, who is also clearly paying attention while dropping seeds into the tray asks, "What about the P, K part of N, P, K? Cover crops are just providing nitrogen. How do organic farmers get P and K and other nutrients?" My answer is, "compost," which is the other important organic fertilizer. "Let's go out to the piles and take a look now."

It takes a remarkably long time for everyone to finish up the last of the sowing, gather together their water bottles and notebooks, and head out to the compost pile, a large, dark brown steaming heap.

"This is compost," I say. "When animal and plant waste is combined in the presence of air, water, and microbes you end up with this dark substance consisting of decayed organic matter and nutrients, some of which are in a form that can be assimilated by plants immediately, and some of which must be further broken down by soil microbes. During the decomposition process, the center of the pile heats up to temperatures between 130°F and 170°F, which kills weed seeds, as well as plant and animal pathogens. Watch David—he is turning the pile now so that all parts of the pile get a turn to heat up in the center."

David is one of the student employees who have been trained to drive the tractor and operate the compost turner. He slowly lowers the large, rotating cylinder carrying the blades into the pile. The tractor starts creeping forward, and the turner throws decomposing organic matter wildly into the metal housing that covers it. The turning process is fun to watch; my students never tire of seeing the escaping water vapor as the composting material gets tossed around. David will turn the pile several times over the next two weeks to ensure that all the material reaches high temperatures and is fully composted.

Once the water vapor dissipates and the tractor goes back to the shop, I continue my talk. "Compost is a good source of N, P, K, and various trace minerals. If you add 10,000 pounds (5 tons) of compost per acre, you will add 100 to 200 pounds of nitrogen, 30 to 150 pounds of phosphorus, and 200 to 300 pounds of potassium. Since these nutrients are chemically bound within organic molecules, the nitrogen will be released gradually, roughly 15% in the first year, with the remainder being released in succeeding years (Van Horn 1995). This is one of the reasons that the transition to organic farming can take a few years. It takes time to build up a supply of organic matter in the soil that will release nutrients at high enough rates to support plant growth".

Recently, at the farm, we tested our soils for nutrient levels. After many years of compost applications, we found that P and K had increased to very high levels. We realized that we needed to emphasize using nitrogen from cover crops and go easy on the compost. We could also have used a variety of pelleted, organic fertilizers made from feather meal, blood meal, chicken manure, or bird guano. Many organic growers use these concentrated fertilizers because they are high in nitrogen (N) and more of the N is available to be used by plants. The pellets are easy to apply and are a useful supplement to cover crops or compost, although they don't add much organic matter to the soil.

There are also other advantages of compost—the microorganisms it contains can suppress soil-borne diseases (Hoitink and Grebus 1994; Lumsden et al. 1983). "Suppress" does not mean eliminate, but compost does help. Compost and cover crops also help to reduce soil erosion. The foliage of cover crops helps protect the soil from the impact of rain drops and wind, and the humus particles in compost are particularly good at holding soil together. On conventional farms, during the off-season, fields are often kept free of weeds by spraying herbicides. In this case, the organic matter in the soil declines, and there is less "glue" to hold soil particles together. The land is then left vulnerable to erosion. Erosion of soil on a farm is similar to erosion of soils from bare hillsides after a big rain. The water simply picks the soil up and moves it, along with any nutrients and pesticides it contains, somewhere else—usually to a place where it is not wanted, like a nearby stream. Erosion leads to a loss of nutrients and topsoil, the sedimentation of rivers, and nutrient and pesticide pollution of waterways. Almost 1.8 billion tons of soil is eroded from U.S. cropland each year (Natural Resources Conservation Service 2006).

At the student farm we make compost both from agricultural wastes and pre-consumer kitchen waste from the coffee house and some of the dining commons on campus. Organic waste, which once went into the landfill, is now composted and recycled back into the farm to provide nutrients for crops.

After putting away the tractor, David returns and tells the group, "There is one thing I have been worrying about. I want to see more organic farms, but as more farms become organic, more organic nitrogen and other nutrients will be needed for fertilizer. What if we can't make enough compost?"

Critics of organic agriculture, like Vaclav Smil, Distinguished Professor at the University of Manitoba in Canada, say that nitrogen from cover crops plus the nitrogen in the manure produced by livestock is inadequate to supply all the fertilizer for the world's farms (Gewin 2004). I answer, "Luckily, no one has to find organic fertilizer for the whole world all at once. The first step would be to return existing animal wastes back to the farm." Our agricultural system is epitomized by centralized, super-sized confined animal feeding operations (CAFO) that produce large, toxic lagoons of animal waste, and it is difficult to return it to farms. The organic agriculture solution to this problem would be to return the animals to the farm so their manure could be composted and used on site. Although UC Davis has cow, horse, pig, sheep, and chicken barns that generate tons of manure for us to make compost, we don't integrate animals into our cropping systems on the student farm. On many other organic farms, however, mixed farming, the integration of crops and animals on a farm, is practiced. In addition to keeping manure on the farm, mixed farms allow animals to feed on crop residues still in the field after harvest. Animals and crops give farmers multiple sources of income and reduce economic risks for the farmer. That said, managing both animals and crops can be challenging and may limit a farmer's vacation time.

Farmers can also apply composted urban green waste for use in agriculture. The state of California mandated that waste going to landfills be reduced by 50%. The key to accomplishing this has been to compost urban, green waste. One successful green waste composter in this area, Grover Landscape Services of Modesto, receives 1,000 tons of green waste each day that is composted and sold to farmers (Goldstein 2005). If green waste compost was used nationwide, it would supplement animal compost and allow more acres of land to be farmed organically.

If after many years, all the animal manure and urban green waste in the country is going to agriculture, and we are still short of fertilizer, then it will be time to update the plumbing in U.S. cities to remove contaminants from sewage so human wastes can be recycled back to the farm. In his book, *Farmers of Forty Centuries*, F. H. King, former chief of the Division of Soil Management of the USDA, described how the return of human wastes to farm lands in China was a model for long-term ecological farming (King 1911). Because sewage sludge in the United States is presently contaminated with cadmium, zinc, and copper, its use in organic agriculture is prohibited by the USDA "organic standards." After all of this talk of manure, sewage sludge, and the buzz of an occasional fly, it is time for lunch. "Anybody hungry?"

Mondays and Thursdays are "pick" days at the farm. From July to October, we harvest sweet corn each week. Today, Sang Min and Jim are picking sweet corn with me, while the other students are picking other veggies for the baskets. I show them how to feel the tip of a mature ear to see if it is filled out. Mostly what they feel is the hollow spot created where a corn earworm has been feeding. The insect deposits its eggs on the corn silk that trails out of each ear of corn. When the larvae hatch, they crawl down the silk into the tip of the ear and begin to feed on the kernels. We open up a couple of ears and see the big, fat, healthy earworms, writhing with irritation at being disturbed from such a luscious feast. The tips of the ears are blackened with "frass," a euphemistic word for insect poop. It is not very appetizing, and our customers deal with it by cutting off the tips. While the first couple of plantings usually have few earworms, the later plantings can get ugly. I view earworms as a problem that both organic farmers and consumers accept in exchange for the benefits of not spraying insecticides, and which may someday be solved in a reasonable way. But Sang Min is not impressed, "You'd better use the pesticide. The pesticide worked well for me." I round up the class once again and we sit down in the shade to talk, putting the work of the farm on hold.

I begin with a description of a large, conventional field planted to a single crop and fertilized with high levels of N, P, K—an optimal place for insect pests to feed and multiply. On these farms, there is plenty of food and little habitat for predators or parasites that prey on pest insects. In such an environment it is not surprising that pest populations can increase rapidly and insecticides are needed. And a tremendous amount of them are used. In California alone, 180,000,000 pounds of pesticides were used in 2004. Despite efforts by scientists and government to reduce pesticide use, the amount used in California increases every year. The Department of Pesticide Regulation has documented a shift towards less toxic pesticides, but the process is slow and millions of pounds of the most toxic materials, like some soil fumigants, are still applied to thousands of acres (Department of Pesticide Regulation News Release 2006). In 2004, the strawberry crop received more than 9.5 million pounds of pesticides, including over 3 million pounds of methyl bromide, a toxic ozone-depleting chemical, banned in many countries (PUR 2004a). We all like strawberries, but the pesticide use seems excessive: more pounds of pesticides were applied to approximately 28,000 acres of strawberries than to 780,000 acres of cotton, and cotton is known to be a pesticide intensive crop! (PUR 2004b). This is largely due to the fact that most conventional strawberry growers don't use crop rotation; they try to grow the very profitable strawberry crop on the same ground year after year. Putting over 9 million pounds of toxins into the environment for high yields of large, red, sometimes sweet, but often not that tasty, fruits doesn't seem worth it. The organic solution is to rotate strawberries with other crops such as broccoli or a cover crop and use disease resistant varieties. With these alternative pest control methods, yields in organic strawberries are 65% to 89% that of conventional production. However, organic strawberries sell for 50% to 100% more than conventional berries, and this higher price helps compensate for the lower yield (US EPA 1996).

"But strawberries are my favorite fruit." Jim says. "Do pesticide sprays matter that much?"

"Good question." I reply, "One study estimates that pesticides cause 10,000 new cases of cancer and kill 70 million birds in the United States each year (Pimentel et al. 1993). Methyl bromide is associated with an increased risk of prostate cancer in farm workers (National Institute of Environmental Health Sciences 2003), and the herbicide, 2, 4D, is associated with a two- to eightfold increase in non-Hodgkin's lymphoma (Zahm and Blair 1992). In a study involving more than 140,000 men and women followed through 2001, those who reported being exposed to pesticides or herbicides before 1982 had a 70% higher rate of Parkinson's disease 10 to 20 years after the initial exposure (Ascherio et al. 2006). So exposure to pesticides in the field can be serious."

"What about pesticide residues on food?" Jim interrupts this listing of cancer studies, "Is there a cancer risk in eating conventionally grown food?"

This is a question on the mind of many consumers. It is known that pesticide residues are found three to five times more often on conventional produce as compared to organic (Baker et al. 2002). Furthermore, some of these residues find their way into our bodies. For example, researchers found that two- to five-year-olds eating a conventional diet had nine times higher than average levels of organophosphate insecticide metabolites in their urine than children consuming mostly organic foods. (Curl et al. 2003). I explain to Jim that despite these numbers, there is no direct evidence that the very low levels of pesticide residues on conventionally grown produce cause harm to human health, and they are usually well below the tolerance levels set by the Environmental Protection Agency (EPA). Still, as a father of two young children, I avoid buying pesticide-treated produce.

I finish this part of my lecture with one more bit of information. "Even with the increasing use of synthetic pesticides we still have as many pests as we did when we started in the 1940s, because pests can evolve resistance to the pesticides. This is because pests that can survive the treatment will grow and multiply. In 1991 researchers documented resistance to pesticides in over 500 insects and mites, 270 weeds, and 150 plant pathogens"(Bellinger 1996).

After these long and depressing descriptions of the toxic effects of synthetic chemicals the class is a little dazed, and it is time for a break. After a little food and drink they return refreshed for a look at how organic farmers control pests.

"Organic farmers use an entirely different and more integrated strategy than the conventional approach of regular applications of pesticides," I explain. "The farmer first needs to learn about the farm ecosystem, including the life cycles of pests and those of beneficial insects that help control the pests. The farmer can then design a farm so that pests are minimized. This starts with crop diversity, but it also means providing habitat for a diverse group of beneficial insects, predatory birds, and mammal predators. Yet diversity alone won't solve pest problems, so the organic farmer also uses genetic control, biological control, cultural controls, and naturally occurring chemicals. For many pests, our naturally occurring bio-control system of lady beetles, lacewings, syrphid flies, various wasp parasitoids, etc. work quite well. So well, that we take it for granted—at least for most crops."

Sang Min has been patiently standing while I talk, with a bag of sweet corn at his feet. He asks, "Why doesn't integrated approach work with sweet corn?"

I answer, "The typical cultural controls, like crop rotation, are ineffective because the corn earworm is not a picky eater and will eat almost any crop that we rotate in such as tomatoes, beans, or lettuce, and the adult moth is a good flyer. I have tried releasing beneficial insects called Trichogramma in the corn but had limited success, and it wasn't worth the cost and effort involved. Breeding for genetic resistance has failed because scientists have not yet been able to find a corn gene that gives protection from earworm."

"Can you spray something organic?" asks Jim.

I reply, "The Bt toxin from *Bacillus thuringiensis* will kill earworms, but the sprayed toxin cannot reach the larvae that have crawled into the ear."

"What is *Bacillus thuringiensis*?" Katie interjects.

"*Bacillus thuringiensis* is a bacteria that produces a toxin (called Bt toxin) that kills a narrow range of moths and butterflies. French farmers first started using *Bacillus thuringiensis* in the 1920s but it wasn't available commercially in France until 1938, and then in the United States in the 1950s. Today *Bacillus thuringiensis* is cultured in industrial production facilities and sold either as liquid or a powder with some additives to make it flow and mix better. After it is combined with water and sprayed in the field, caterpillars eat the bacteria in the form of spores and toxin. The toxin destroys the gut walls of the caterpillars and spores and other gut bacteria invade its body. This approach is an example of 'biological control,' using live organisms to combat pests and disease.".

Unfortunately, Bt toxin doesn't work very well for the corn earworm making it difficult to control this pest on organic farms. Researchers continue to work on this problem, and hopefully a more effective organic solution will be found. I recently learned that injecting vegetable oil and Bt toxin into the silks of each ear with a special applicator was very effective and not overly time-consuming. I can't wait to try this technology (Diver et al. 2001).

On our way back to the packing shed, we walk over to a bare spot on the farm. This wasteland is the result of another pest problem. Nothing grows here because of symphylans, an arthropod that is closely related to centipedes and millipedes. These white, quarter-inch-long, 24-legged pests have proved difficult to control organically. Usually, the first indication of a symphylan infestation is a relatively small area of stunted, unhealthy plants. Sometimes, the entire crop in the infested area will be completely destroyed. Each year, the affected area increases in size by about ten to twenty feet. At the student farm we have tried controlling symphylans with flooding, non-host crop rotations, letting fields lie fallow, and mechanically disturbing the soil, but nothing has worked well. How do conventional growers control this pest? They use very toxic materials like methyl bromide. Clearly new strategies are needed for both organic and conventional growers.

<hr/>

Later in July, on what seems to be an equally hot day, the class is hoeing melons. The beds, which were planted early in the spring, have not been solarized and are weedier than I would have liked. Sang Min uses this time to share his experience with growing melons. He points at the melons, which have begun to spread their vines across the bed, and says, "Not best way to grow melon. You must prune melon to four stems, and remove all flowers for forty centimeters along each vine. Otherwise, bad fruit."

"Hmmm," I respond, "I have never seen anybody do that."

He continues, "Plant is like young girl. You don't want plant pregnant when too young. Fruit is not good, too small."

"Really," I reply, "no one around here does this and lots of people grow melons."

Sang Min doesn't give up. "Fruit is bigger and harvested all at once if you prune. We pick four containers per acre. Good size, everyone buys our melons."

I yield to the melon master. "Show us, Sang Min. Show us how to prune the melons." He bends down, selects a vine, and explains how to pick off flowers and select vines. Once we understand, we start pruning flowers and do half a row as an experiment.

When we finish the pruning, there are still three more melon beds to hoe. We occasionally have students at the farm who are very enthusiastic about hoeing. I don't know what motivates them. For me, hoeing is necessary sometimes, and I can go into "small mind" mode and get it done, but I don't look forward to doing it. Yet, like sowing seeds in plug trays, it is a great time to talk, get to know each other, and answer questions. As we continue cleaning up the melon beds, Sang Min asks, "What if prices of organic food too high for consumer?"

Sang Min isn't so interested in the hands-on parts of our class; he wants to sell a lot of produce and make money. "It is true that many of the organic products are more expensive than the conventionally grown," I answer. "In June, at the Davis Food Co-op organic apples were $2.99/lb, while conventional apples were $1.19/lb. Organic carrots were $0.99/lb, while conventional carrots were $0.59/lb. Whatever the price, many consumers in the United States and Europe are willing to pay it. They are paying for a cleaner environment because fewer synthetic pesticides and fertilizers are contaminating land and water. They are paying for greatly reduced pesticide residues on produce and for produce that may have higher levels of antioxidants and other nutrients (Asami et al. 2003). They are paying for better health, and a few preliminary studies suggest that they may be getting it (Asami et al. 2003). Given the environmental benefits, health benefits to farm workers, and possible nutritional benefits, higher priced organic produce is still a bargain. It is the unpaid environmental costs of toxic pesticides and synthetic fertilizers that may end up being too high a cost in the long run."

I go on to explain that many small organic farm businesses in California are financially successful because they market directly to consumers, either at farmer's markets, through subscription services (called CSA for community supported agriculture) or farm stands. Direct marketing eliminates middlemen, which means the farmer sees more of the sales dollar. Consumers can also benefit from direct marketing. Our direct market prices, when I worked at Full Belly Farm and now at the student farm, are lower than retail prices. Also, the farmer determines the price in direct sales. An organic farmer who farms near Davis had an employee who went on to get an M.B.A. As part of her degree, she analyzed the net return on produce sales from the farm. She found that both the CSA and farmer's market sales had a net return of about 40%, that the return on retail sales was 3.5%, and that wholesalers were losing money. (Legarre et al. 2001). I hope that Sang Min recognizes the additional benefits of direct marketing: building a local community that supports the farm and the farmers. I can see him fitting in well at the Farmer's Market, cajoling and admonishing his customers to buy his beautiful melons, but it is going to be tough to convince him to sell in farmers markets.

Fortunately for Sang Min, it seems that for larger operations there are profitable opportunities for selling wholesale. I suggest that Sang Min check out a local Davis supermarket's organic section. There he is likely to find produce, milk products, and a wide variety of processed products like cereal, pasta sauce, breads, and desserts. Processed products in particular have fueled the phenomenal growth of the organic industry. Within the past fifteen years, supermarkets in California went from having almost no organic products to devoting large sections of their shelves to organic products. Large farms are now making profits wholesaling to large corporations, who increasingly want to share in this profitable sector of agriculture. At the corporate level there has been considerable consolidation in the organic foods industry. Many of the familiar "organic" labels are now owned by large corporations (Weeks 2006). For example, Kraft Foods owns Boca Burger Inc., and, in 2004, bought the natural cereals producer, Back to Nature. Kraft is a subsidiary of Altria Group, which also owns Phillip Morris Companies Inc.—one of the largest cigarette makers in the world. This is just one example of at least a dozen corporations buying smaller organic food companies.

Now, even Wal-Mart, one of the nation's largest grocery retailers, is selling organic products and vowing to make them accessible to lower- and middle-income America (Warner 2006). With the greening of Wal-Mart, demand for organic produce is expected to increase even more dramatically—and it has already been growing at 15% or more a year for the last ten years. On one hand, this decision will help expand the amount of land that is farmed organically, reducing soil erosion and pesticide and synthetic fertilizer use and sparing wildlife. On the other hand, some worry that this expansion could lead to a dilution of the meanings and practice of organic agriculture. For example, Julia Guthman, a professor in the Department of Community Studies at the University of California, Santa Cruz argues that the high-value crops and the lucrative segments of organic commodity chains are being appropriated by agribusiness firms, many of which are abandoning the agronomic and marketing practices associated with organic agriculture (Guthman 2004). Author and journalist Michael Pollan addresses this question in his best selling book *The Omnivore's Dilemma*. After visiting Earthbound Farm, a giant grower that sells much of the fresh organic salad mix and spinach in California he said, "I began to feel that I no longer understood what this word I'd been following across the country and the decades really meant—I mean, of course, the word 'organic' . . . [in] precisely what sense can that box of salad on sale in the Whole Foods three thousand miles and five days away from this place truly be said to be organic? And if that well-traveled plastic box deserves that designation should we then perhaps be looking for another word to describe the much shorter and much less industrial food chain that the first users of the word organic had in mind?" (Pollan 2006).

I see it more pragmatically. In the 1980s and early 1990s, when organic farmers and activists were contemplating what organic farming "success" would look like, most agreed it would be when mainstream supermarkets were all offering organic vegetables and processed products. This scenario would mean that organic agriculture had changed the world, and it would be a better place. If small organic growers cannot supply supermarkets with what they need, the markets will buy from larger, industrial operations. Large growers must follow the same organic standards as everyone, so there is an overall environmental benefit. The organic food system has two production and sales models operating simultaneously—the small, local and the large industrial. The first serves the people who want to buy from farmers' markets, CSAs, or "u-pick" operations while the other serves the majority of people who shop in supermarkets.

If the presence of these large growers lowers wholesale prices, it makes it more difficult for smaller growers to succeed in the wholesale market. I don't see that as catastrophic; first, there will always be a place for smaller growers who direct-market, because many consumers want to buy locally to support their community and get the freshest produce possible. Second, marketers like Whole Foods recognize that local is part of the ethic of selling organic food. On a trip to visit Pam's brother in Mill Valley, CA, I walked into the Whole Foods store there, and was surprised to see the numerous "Locally Grown" signs throughout the store. "Local" in this case meant about a 100-mile radius, but that is much better that the 1,000 to 1,500 miles industrial produce usually travels.

Sang Min's farming operations are mostly in Mexico. The market there for local, organic produce is limited. If he wants to go organic, he needs to enter the increasingly competitive organic wholesale market in the United States. However, with lower labor and land costs in Mexico, he just might be able to make it. The UC Davis agricultural economist, Steve Blank, predicts that most agricultural production in the United States is going to move overseas, just like our manufacturing has moved (Blank, S. C. 1998). Perhaps Sang Min is the model organic farmer of the future! Nothing against him personally, but I'd rather buy local.

—⟨⟨⟩⟩—

My back is getting tired from hoeing and it's only getting hotter as the morning wears on. I am ready for the questions to stop so we can stop working, but there is one more. "Does organic farming yield as much as conventional farming?"

This is not an easy question to answer, because it depends on the crop, the place, the farmer, the variety, the type of crop rotation used, and whether cover crops take the place of crops. Bill Liebhardt, the director of the Sustainable Agriculture Research and Education Program at UC Davis during the 1990s, determined that organic corn yields were 94% of conventional corn, organic soybean yields were 94% of conventional, and organic wheat yields were 97% of conventional. In processing tomatoes, based on 14 years of comparative research at UC Davis he found no differences in yield (Liebhardt 2001). Other researchers have found different results. A European study found that with a three-crop rotation of wheat, potatoes, and grass/clover, organic yields of wheat were 90% of conventional; organic potato yields were 58% to 66% of conventional, and organic grass/clover yields were similar to the conventional (Maeder et al. 2002). At the long-term research farm at UC Davis, after nine years of growing a corn and tomato rotation, organic corn yields were 66% of conventional, while there was no statistical difference between organic and conventional tomatoes (Denison et al. 2004).

Although it is possible to criticize these studies for a variety of reasons, one common factor I observe is that the organic systems studied often had less mineralized nitrogen available. Therefore, crops that have a strong yield-response to nitrogen, like corn, may do better in conventional systems all other things being equal. Situations do arise when the effect of nitrogen is overshadowed by a different growing factor. For example, during drought years at the Rodale research farm, the organic corn out-yielded the conventional corn (Pimentel et al. 2005) Nevertheless, the real question may be whether an increase in yield justifies the environmental damage caused by increased fertilizer and pesticide pollution.

For the crops I have grown or inspected here in California, yields on organic farms were comparable to conventional farms if the organic farmer (including me) did a good job controlling weeds, supplying adequate nutrients, planting similar varieties, and didn't have any serious pest problems. This is not always the case.

Organic rice grown in California often has significant weed problems that can reduce yield 30% to 50%. Organic rice growers tolerate these lower yields because their inputs for weed, insect, and disease control are very low compared to conventional systems. Net profit is usually higher in the organic system because the prices for organic rice have stayed high, while conventional rice prices have barely risen above the costs of production. Still, better ways to control weeds would help make organic rice yields more comparable to conventional rice, which, as the world's population increases, may be an increasingly important issue.

As Sang Min brought up earlier, there are times when cover crops may be grown in place of food crops. When this happens, short-term reductions in yield are exchanged for longer-term sustainability.

The students stop weeding and look up at me, trying to absorb the complexities of yield comparisons and I feel the need to sum up the issue cleanly. The best I can do is to tell them that skilled farmers, using best organic practices and technologies, can achieve high yields while caring for the environment. If the goal of this course is give these students the knowledge and skills to do this, we have a ways to go.

We finish hoeing the melons, and the plants look a little beaten up by the process. After a few days they will have grown enough new leaves and vines to hide any evidence of our morning's hot work.

<hr />

Weeks later, near the end of summer and the course, we inspect our melon pruning handiwork. Sang Min's melons are large and luscious (and a little later maturing than the ones we didn't prune). We share them and I thank the students for taking time to learn and for sharing their knowledge with me. These students will end up in a variety of places. Some have become organic farmers and others work overseas doing agricultural development work. Some work for food banks, state or federal agricultural agencies, or at the local food co-op. They are all contributing to a more sustainable agriculture system.

A few weeks after the class is over, at the beginning of September, I take the clear plastic off the solarized beds. Considerable planning, resources, and time have gone into eliminating weeds, so I plant the beds as densely as possible. I load carrot seeds into the student farm's two Stanhay belt seeder units that have special "triple shoes" and put beet seed into a single row unit in the middle. In each bed I plant six rows of carrots along with a row of beets; I plant four beds, thinking that 46,000 carrots and almost 8,000 beets should be enough for the sixty baskets of produce we sell each week. I start the sprinklers right after I plant. They run six hours today and then a couple of hours every other day until the carrots and beets germinate. After ten days the seeds have germinated and the stand looks pretty good. We continue to irrigate weekly until the fall rains kick in. Before Thanksgiving, we start to harvest our first carrots. They are crisp, sweet, and well formed. We haven't spent a minute weeding the beds.

Three

THE TOOLS OF ORGANIC AGRICULTURE

A truly extraordinary variety of alternatives to the chemical control of insects is available. Some are already in use and have achieved brilliant success. Others are in the stage of laboratory testing. Still others are little more than ideas in the minds of imaginative scientists, waiting for the opportunity to put them to the test. All have this in common: they are *biological* solutions, based on understanding of the living organisms they seek to control, and of the whole fabric of life to which these organisms belong. Specialists representing various areas of the vast field of biology are contributing—entomologists, pathologists, geneticists, physiologists, biochemists, ecologists—all pouring their knowledge and their creative inspirations into the formation of a new science of biotic controls.

RACHEL CARSON, *Silent Spring*, 1962, p. 278

In the half dark of early morning I put on my pants, getting ready to go to work. Pam peers across the bedroom and her not-so-sharp eyes somehow instantly notice something is amiss. "Is that a hole in your pants?" I look down. There bulging from my left thigh, the Swiss Army knife in my pocket has indeed worn a hole through the pant leg. I'm not surprised. The knife comes with me wherever I go.

As a toolmaker and tool user considerable thought goes into choosing which Swiss Army knife works best for me. At a minimum the knife must contain: scissors, a corkscrew, tweezers, a screwdriver, a saw, a bottle opener, and of course, a couple of blades. The Huntsman Plus fits the bill with the least weight. Once I had a Swiss Champ. It has everything mentioned above, plus a magnifying glass, pliers, pen, and thirty other features. But people usually laugh when you bring it out. I get by with the Huntsman Plus and only occasionally wish for the pliers and pen.

I walk down the hall and into the kitchen, headed for breakfast. I don't know who or what the cereal companies think they are keeping out of the cereal, but most adult men are incapable of opening a sealed bag of cereal. Super guys can do it, but the bag usually explodes all over the room. When I was younger I pulled and tore. Now, I use my knife's scissors.

Belly full, tea in hand, I head to the door. Kisses and hugs to Pam and the kids who have arisen and are shining. I don't get out the door though. Cliff, six, has a soccer game in the afternoon and wants me to help find and put on his uniform. But first he holds out his fingers. "Daddy, the coach says I have to cut my nails or I can't play." I grab his jittery little fingers and snip, snip, snap with the knife's scissors. He holds surprisingly still and there is no blood. Meanwhile, Audrey, four, has started crying. She is getting dressed, and her pants have no belt. The belt that I find in the closet, given to her by her older cousin, is too long—about eight inches too long. With the awl on the knife, I make new holes for her four-year-old waist, and with another snip, snip, snap, I shorten the belt to fit. Audrey is mollified, and I am late.

I arrive at the farm, open the doors of the packing shed and set thirty-one empty baskets for our subscription produce service on the large table. I can see the dry-erase board on the wall with the list of produce we will pick: Red Russian Kale, spinach, turnips, escarole, butternut squash, sweet potatoes, eggplant, peppers, arugula, lettuce, and a pumpkin. Early fall is a bountiful time of year, distinguished by the overlap of warm-season vegetables with cool-season ones.

Students begin to arrive and I direct them to the harvest list. "Hey Derek, how about if you and Morgan pick the kale this morning? We have thirty-one baskets to fill today. Katherine, do you want to start the sweet potatoes?" I have something else on my mind: those two beds of late-planted melons sown by Sang Min and the other students from my summer class. They have been ripening very slowly, but it would be such a sweet surprise for the shareholders to receive a melon in October. I walk to the beds and look at the Sivan Cantaloupes. Earlier in the year the melons in the first planting were large and delicious. Now they are smaller, but plentiful, and some are turning a ripe golden color. I choose a likely candidate and pull the fruit away from the vine. The place where the vine connects to the fruit slips easily. The fruit should be ripe. I open the largest blade on my knife and slice deeply into the melon, cutting quickly in a smooth motion. Sweet melon nectar oozes out onto the knife. I cut again, carving a slice from the melon, and then another and another. I hear a laugh and look up to see that I am surrounded. It seems that students are mysteriously drawn to melons. I pass out slices and then save one for myself. The flesh is vividly orange in the morning light, and as the fruit moves toward my mouth I can remember the taste of many wonderful melons harvested earlier in the summer. I bite into the cold, juicy flesh and taste the sweet, musky melon-ness, but then, the over-musky, chemical terpene, bitter aftertaste of bad, poorly ripened melon. We taste a few more with the same result. "Oh well, no melon surprise today."

I head off to pick spinach. The plants are at perfect maturity. With my knife I cut the base of the Winter Bloomsdale Savoy Spinach with one hand, while holding the tops of the leaves with the other. Great handfuls of spinach fall into the box; a rain of dark green, crinkled leaves, full of life and covered with dew.

I am finished picking and head over to check on the red kale and lettuce. They are drooping a bit after being hit with two days of fierce north winds so I set up a line of sprinkler pipe to irrigate them. I open the valve to release water into the sprinkler pipes. The water surges into the pipes and, one after the another, the nozzles along the line spit out a little water, until with a "whoosh!" the line pressurizes and long arcs of water pulse out over the crops. Sprinkler pipes are the naughty children of farm equipment, always getting into trouble, and sure enough the last sprinkler nozzle is plugged with something. I poke the awl on my knife into the nozzle, trying to dislodge what's in there. Poke and push, poke and push. Finally, the mini-dam breaks and water rockets into the air. I pause to admire the uniform pattern of water rising over and falling on the lush greens. The purplish reds of kales, the dark green spinach leaves, the paler, frilly greens of lettuce, all sparkling with splashing water droplets in the sunlight, are a very beautiful sight.

I recall a wedding I attended last month in Tucson to celebrate a couple who had met at the farm and had been two of my favorite students and friends. I left straight from the farm, changing my clothes in the shop, transferring the contents of work pants' pockets to wedding pants' pockets: keys, wallet, knife—my daily triumvirate. I hurried to the car, and eating a bag lunch, sped to the airport.

I don't like airports. They are portals to another world, a world of high technology, amorphous crowds of bustling people, bad food, bad air, no plants, and a risk of delay if not disaster. I always feel like an alien. With my carry-on bag in tow, and e-ticket in hand, I checked in at an interactive kiosk and headed to security. The line in Sacramento was short and I moved quickly through to the metal detector and emptied my pockets into a plastic tray. Oops! I realized that I had my Swiss Army knife in my pocket. It couldn't go on the plane as a carry-on. I paused and thought. I could surrender it to the Department of Homeland Security, or I could go back and check my bag and put the knife in it. I pondered the risk of lost luggage and the purgatory of baggage claim, but it was a simple decision. This knife connects me to food and family and work, and it would not be easily surrendered. I would need it in Tucson to hack my way through the wilderness and return to home safely. I sheepishly backed away from security—no fast moves—and checked my bag.

The mechanical technology we use at the student farm and what most organic farmers in this area use is hardly any more sophisticated than my Swiss Army knife, and certainly less fancy than the technology used in modern air travel. Tractors, trucks, mowers, discs, spaders, Lilliston rolling cultivators, cultivating sleds, compost spreaders, irrigation pumps and pipes, drip irrigation equipment are all twentieth-century, iron and steel technology with a little bit of plastic thrown in. With few exceptions, it is the same equipment that is used in conventional agriculture. The technology that sets organic agriculture apart from conventional agriculture isn't mechanical; it's ecological and biological.

From preparation of seed-flat mix to pest control, organic agriculture follows its own paradigm. It uses the technology of the living. Our seed-flat mix is a combination of aerobic compost, peat moss, and vermiculite. The aerobic compost is full of microorganisms that have been shown to have a suppressive effect on plant diseases. Instead of sterilizing our mix, as is done in conventional agriculture, organic farmers let the beneficial fungi and bacteria and yeasts in the compost out-compete the disease-causing organisms. From what I can tell after twenty years of organic greenhouse seedling production, it works.

The same principle applies to insect pests in the greenhouse. The student farm has the only certified organic greenhouse on campus. Like most greenhouses, we have pest problems: fungus gnats, white flies, and aphids. Instead of applying toxic pesticides that try to kill everything, we release beneficial insects and nematodes to manage these pests. One researcher on campus, Mike Parrella, has been working on biocontrol in greenhouses for many years; he, his staff, and students are trying to expand this biocontrol technology to the rest of the greenhouses on campus. This year (2005) Parrella's group provided us with a control for the fungus gnats that had been stunting and killing our seedlings. We mixed fifty million *Steinernema feltiae*, a predatory nematode, with water and sprayed them on the plants. The nematodes crawl down into the seed-flat mix to the roots and attack the gnat larvae through the mouth and anus. They then release bacteria into the gut of the insect, which feeds on and digests the larvae. The nematode then eats the bacteria.

This paradigm of using living organisms is also applied in the field. We add compost to soil in part to add beneficial microorganisms to the soil. We enhance our beneficial insect population by planting borders with flowering perennials that provide them with nectar and pollen, and by growing a diverse selection of plants in the garden, many of which also support beneficial insects. At the farm we set up owl boxes to attract owls to control rodents, and try to provide structures and habitat that will support other raptor birds. Farming organically, using biological technology to farm, instead of chemical technology, results in diverse, living farms that support raptors and other predators, beneficial insects, soil microbes, worms, and flowering plants. It creates a farm that functions as a whole ecosystem with many interactions. It is a healthy place to grow plants, to work, and to visit.

***

Where did the ideas of organic agriculture originate? Modern organic farming began with Sir Albert Howard, a British agriculturalist working in India in the early 1900s. Howard believed that the supreme farmer was Nature, and was very critical of synthetic fertilizers, believing that they would destroy soil structure and quality. He also made a direct link between health of the soil and health of the plant and health of the people, and observed that civilizations in the past had collapsed due to poor farming practices. He wrote two books: *An Agricultural Testament*, and *The Soil and Health: A*

Study of Organic Agriculture, which were required reading in my first organic farming class (Howard 1940, 1947). A paragraph at the beginning of *An Agricultural Testament* summarizes his views:

> Mother earth never attempts to farm without live stock, she always raises mixed crops; great pains are taken to preserve the soil and to prevent erosion; the mixed vegetable and animal wastes are converted into humus; there is no waste; the processes of growth and the processes of decay balance one another; ample provision is made to maintain large reserves of fertility; the greatest care is taken to store the rainfall; both plants and animals are left to protect themselves against disease.

He was impressed with traditional Indian farming practices and helped to develop and disseminate the making of compost piles from animal and vegetative waste, as well as the use of leguminous cover crops to improve soil fertility.

In Germany during the same period, Rudolf Steiner developed another system of organic agriculture called Biodynamics in response to local farmers' perception that soils were becoming depleted when chemical fertilizers were used. Steiner's organic system emphasizes spirituality over peer-reviewed science, and is concerned mainly with life-forces and the vitality of the earth. Biodynamics tries to achieve a balance between physical and higher, nonphysical realms. It relies on a mix of animals and crops, compost, and a number of "preparations" made from herbs and manures to create a system intended to yield well and promote health, with very few off-farm inputs. Today, Biodynamics is more popular in Europe than here. In the United States, "biodynamic" is not necessarily the same as "organic" because it does not comply exactly with USDA organic standards.

In 1947, J. I. Rodale, an American who was influenced by Sir Albert Howard, established an experimental organic farm in Pennsylvania and published organic gardening and farming magazines and books. The Rodale book, *How to Grow Vegetables and Fruits by the Organic Method*, has been an important source of information for organic growers for many years (Rodale 1976). At the time it was published, the editors relied more on anecdote than on science, but (I suspect) were hoping other scientists and their own experimental farm would eventually validate the anecdote.

Rachel Carson had a crucial impact on the expansion and acceptance of organic farming. She published *Silent Spring* in 1962. This seminal and best-selling book explained how the widespread use of pesticides in agriculture damaged the environment. *Silent Spring* triggered consumer awareness and a demand for produce that did not contain pesticide residues and a farming system that did not destroy the environment.

For someone who didn't write any books, and whose greatest fame in other parts of the world was as a Shakespearean actor, Alan Chadwick had a strong influence on organic farmers in California. Alan was invited by the UC Santa Cruz Chancellor,

Dean McHenry, in 1967 to set up the Student Garden Project on campus (Hagege 2003). Chadwick, who had been raised in an upper-class Victorian English household, brought with him French Intensive, Biodynamic organic gardening techniques. His French Intensive knowledge came from English and French market gardens that operated during the first half of the century. Double digging, composting, cover crops, cold frames, vegetable husbandry, Chadwick knew it all, and had an ability to create beauty wherever he went. His Biodynamic knowledge came directly from the source; Rudolf Steiner had once been his tutor. By combining the two, Chadwick had a system of techniques and spirit that appealed to students in the sixties. He was a charismatic and inspiring teacher, but also had a bad temper that alienated many people. While he was working at UC Santa Cruz, he was a focal point for the conflict between mysticism and science in organic agriculture that still exists today. The UC Santa Cruz Farm and Garden Apprenticeship is the ongoing legacy of Chadwick. Scientific rigor has been applied to his teachings to create a curriculum that transformed hundreds of students from around the country into organic farmers.

This brief history of organic agriculture suggests that organic "technology" is derived from a combination of a Brit's vision of nature as the supreme farmer along with a translation of Indian farming traditions, a spiritual philosopher's insights into the natural world, a Shakespearean actor's gardening hobby, and an ecologist's outcry against environmental destruction. Have research scientists contributed anything to the success of organic farming?

The answer is yes; despite the fact that Sir Albert Howard had a dismal view of science, and thought that agriculture should develop in the field, not in the lab. Take, for example, the discovery of *Bacillus thuringiensis*, the most commonly used organic insecticide. In 1901, the Japanese biologist Shigetane Ishiwatari was investigating a disease of silkworm larvae, and identified the bacteria *Bacillus thuringiensis* as the agent that caused the death of these economically important insects. In 1911, the German scientist, Ernst Berliner rediscovered *Bacillus thuringiensis* while trying to figure out what was killing Mediterranean flour moths. It wasn't until 1956 that the researchers Hannay, Fitz-James, and Angus found that the main insecticidal activity against moth insects was from a crystal endotoxin (BT) produced by the bacteria. This discovery paved the way for using Bt toxin as a pesticide, and it became commercially available in 1958 (Aronian Lab 2006).

While *Bacillus thuringiensis* is an example of biological control of insects using bacteria, there have also been scientific advances in the biocontrol of insects using other insects. Charles Valentine Riley, the Chief Entomologist for the USDA, began importing beneficial predators and parasites into the United States beginning in 1873. The most famous of the introductions was the importation of an Australian lady beetle to California to control the cottony cushion scale on citrus. After this very successful biocontrol importation, hundreds of parasites and predators were introduced into the United States to control exotic insect, mite, and weed pests. In 1923, the University of California established what would become the Division of Biological Control to research and develop biological control solutions for insects, mites, and weeds. While there were failures as well as successes at the Division of Biological Control, much knowledge about pests, predators, and parasites has been established and many students were trained to further develop these techniques.

Another important breakthrough for organic agriculture came with the discovery in 1870 by Jean-Henri Fabré of insect pheromones, chemical substances that help insects communicate with each other. It was not until 1967 though, that Harry Shorey at UC Riverside found a way to use pheromones for mating disruption in cabbage looper moths. This principle was extended to control other insect pests like the codling moth that infects apples. When distributed throughout a field or orchard, the mating pheromone of the codling moth confuses the males so much they can't find the females to mate. No mating, no larvae. No larvae, no pest damage. Growing marketable organic apples without the pheromone mating-disruption of the codling moth would be very difficult.

Plant breeding for resistance to disease began in England over 100 years ago. In 1904 commercial wheat was susceptible to a fungus that caused a disease called stripe rust. Small, yellow, elongated pustules would appear in rows on the leaf, eventually forming long, narrow, yellow stripes. When the pustules matured, they would break open to release a yellow-orange mass of spores. By crossing a non-commercial resistant variety with the commercial variety, a resistant variety was generated that had the qualities of both parents (Chrispeels and Sadava 1994). This began an extensive practice among plant breeders of introducing pest and pathogen resistance genes into plants. Today most of the crops we eat contain such disease resistance genes.

Breeding for disease resistance is usually not a permanent solution. Pests and disease-causing organisms can evolve rapidly to overcome the resistance, and the plants eventually become diseased again. It keeps breeders in business, however, because as new resistance traits and their genes are found, they can be bred back into plants. This genetic approach of introducing disease-resistance genes into cultivated crops has been the mainstay of agriculture for the last 100 years, and is one of the technologies that make organic agriculture possible.

Over the years, scientific research has validated and extended organic principles. Even though organic farming has been successfully practiced around the United States since the beginning of the century, by 1980 there had been very little research solely for organic agriculture. According to Patrick Madden, the first director of the USDA's LISA program, many agricultural scientists and conventional farmers, thought organic agriculture lacked credibility until the scientific community in the United States issued three important reports. The first was the United States Department of Agriculture's (USDA) *Report and Recommendations on Organic Farming* (USDA 1980), which gathered and reviewed scientific evidence on yield and net return of organic farming, and made recommendations for research, education, and public policy. The authors visited and studied successful organic farms throughout the United States. They found farms that were making compost from animal waste, rotating crops, planting cover crops, and growing crops successfully and profitably—without the benefit of much scientific research. This report was ordered by Secretary of Agriculture Bob Bergland of the Carter Administration. Despite the very positive conclusions of the report, its recommendations were thrown out by the incoming Reagan USDA. The second report, entitled *Alternative Agriculture*, was from the National Academy of Sciences. It summarized the current scientific knowledge about tillage, biological control, and cover crops as a source of nitrogen, as well as detailed the problems in conventional agriculture caused by pesticides (Board on Agriculture 1989). In 1990, the General Accounting Office (GAO) issued the third influential report, which documented widespread public concern about the detrimental effects of pesticides on the environment, human health, and the quality of life (U.S. GAO 1990). With the momentum generated by these reports and increasing demand by the public for an alternative to conventional agriculture, organic growers and activists began to lobby the USDA for increased funds for organic research. Their efforts led first to the Low Input Sustainable Agriculture (LISA) Program that began in 1988, and which eventually evolved into the Sustainable Agriculture Research and Education (SARE) Program (Madden 1990). From 1988 to 2006 the SARE program provided $214.5 million for sustainable agriculture research, education, and extension, funding over 3300 projects (Jill Auburn, personal communication). SARE provides information to farmers on improving sustainable agricultural farming practices, based on the results of this research.

Soon after the LISA program began, Bob Scowcroft started the Organic Farming Research Foundation (OFRF) to raise money from private foundations to fund research on organic farming. Since 1992, OFRF's grant making program has awarded more than $1.5 million for over 200 projects. Over this period, 67% of the funds have gone to professional (university-based) researchers, 15% to farmers, and 18% to nonprofit organizations. Their grant-making objective is to generate practical, science-based knowledge to support modern organic farming systems. OFRF-funded projects have contributed researched-based information to many areas of organic agriculture. For example, they have funded projects that tested the effectiveness of beneficial insects in a variety of crops, bred crop varieties for organic systems, evaluated the effectiveness of birds and bats on insect populations, and analyzed the food quality benefits of organic products. (Sooby 2006).

These and other organic research efforts, although not huge in terms of dollars, have significantly advanced and refined organic agriculture. Some practices have been validated and others discarded, new techniques have been developed. More importantly, the increasing involvement of university scientists has ensured that organic production practices are science-based and effective.

Some science and development has also come from private agricultural companies. For example, Agraquest, has developed biologically-based products like Serenade

(Agraquest 2006), which can control powdery mildew on grapes and is a good substitute for sulfur. Drs. Michael Glenn and Gary Puterka of the USDA Agricultural Research Service (ARS) in Kearneysville, WV, working with the Engelhard Corporation, developed a product called Surround. Made of kaolin clay, it is sprayed on pears and apples to control leafroller, and leafhoppers, and suppress mites, codling moth, apple maggot, and other pests. First marketed in 1999, this product is now widely used in organic fruit production (ATTRA 2005). Interestingly, these two products are used on both organic and conventional farms. Environmentally sound, low toxicity pest control materials are useful to all types of growers and are an indication of how large an impact science-based, organic agriculture research is having on agriculture in general.

What is the future of organic agriculture? Based on present trends, consumer demand and organic-based research are likely to increase. The question is whether the technology of organic agriculture is robust enough to meet the growing demand for food and fiber around the world. Organic agriculture can certainly help address the problems of environmental degradation associated with conventional agriculture, yet some critics suggest that a third more land would need to be farmed if all the agricultural land in the United States was farmed organically (Trewevas 2006). Whether this is accurate is debatable, but it is clear that for organic agriculture to be successful in feeding the world, huge changes will be needed: recycling of organic waste back to farms for nutrients, development of crop varieties with enhanced tolerance to pests and stresses, and reduced meat consumption so that more of the food crops can go to humans rather than animals. I don't know how quickly these changes could be made, or what level of change would be socially acceptable and economically feasible. What I do know is that as an organic farmer, I want to see more farmland transitioned to organic practices and at the same time I want to use the most powerful technologies available to create an environmentally friendly, sustainable, and high-yielding farm.

This raises the question of whether GE varieties can help forge a future sustainable agriculture that can meet our criteria. Is GE part of "better living though biology" or "better living through chemistry" or something else altogether? It is worth asking this question because GE has the potential to increase resistance of plants to insects, diseases, and nematodes, and help plants adapt to environmental stresses like drought, flooding, cold, and salt. In the same way that the introduction of genes from wild species through breeding revolutionized farmers' management of pests, so can the introduction of genes through GE revolutionize control of diseases, insects, and nematodes for which there is presently no organic solution.

GE can also greatly increase our understanding of what is going on in plants at a molecular level. Pam has been working for twenty years trying to understand how plants and microbes communicate. Why are some plants resistant to disease, and others not? How do disease-causing organisms break down these defenses? What does the plant do to defend itself? Clearly this dance is governed by biology, that is, proteins and genes. What are they and how do they function? Maybe if Rudolf Steiner was alive today, he would be in a lab introducing viral gene snippets into plants to make them immune to viruses—a sort of biodynamic, homeopathic cure.

At present, though, the door to using GE plants in organic agriculture is firmly closed. When the USDA was drafting the national organic program standards, it came out with a version that included the use of GE plants. Judging by the 275,000 overwhelmingly negative letters sent by the organic community, the plan was unthinkable. GE plants, therefore, were excluded from the National Organic Program (NOP) standards that were implemented in 2001 (USDA 2007). Today organic activist groups like the Organic Consumer's Association, California Certified Organic Farmers, Californians for a GE-Free California, and many others continue to lobby against GE plants. Why is this?

Many in the organic community view GE as a synthetic process. J.I. Rodale, one of founders of organic agriculture, believed that "If it is synthetic, avoid it. If it goes through a factory, examine it with special care. Follow the dictates of the cycle of life when growing things and you will be blessed with foods of surpassing taste and quality that are less troubled by insects or diseases" (Kupfer 2001). This fairly fundamentalist view implies that there is a natural way of farming that can be rediscovered like a Garden of Eden. Fundamentalists would be particularly opposed to any GE plants that contain genes from bacteria, like putting Bt toxin into corn.

Others may have concerns that are more scientific. There has been much concern about the spread of pollen from GE plants and how this movement could affect non-GE crops and native plants. If organic crops cross-pollinate with GE crops, it is possible that consumers would reject the crops. There is also the concern that GE traits could be transferred and persist in wild plants in such a way as to disrupt the natural ecology. I can imagine that Rachel Carson would have been concerned about the potential disruptive effects of pollen flow. At the same time, she may have thought GE plants could be beneficial if they could help reduce pesticide use.

Quite naturally, too, people are worried about food safety. Is it safe to eat transgenes? Is it safe to eat the products of the transgene such as Bt toxin? Alan Chadwick would also ask, "Does it taste good?" One of the reasons the first commercially available GE plant, the Flavr-Savr tomato was a failure, was that, despite the fact that it had extended shelf-life, it had no more flavor to savor than any other prematurely picked, industrial tomato. Even if GE plants are safe to eat (See chapter 7), are they worth eating?

Finally, there is fear that this technology is owned, and will continue to be owned, by large corporations. It is distressing to think that something as magical as seed will cost a lot of money and can no longer be propagated by the farmer, but that exactly describes hybrid seed as well as some GE seed (See chapter 9).

Many organic farmers and organic consumers don't know how genetic engineering of plants actually works. I can explain Swiss Army knives, and even organic technology, to anyone who cares to stand next to me and hoe melons, but I will leave the description of genetic engineering to Pam.

The Lab

Four

THE TOOLS OF GENETIC ENGINEERING

Wisdom demands a new orientation of science and technology towards the
organic, the gentle, the non-violent, the elegant and beautiful.

E. F. SCHUMACHER, *Small Is Beautiful*, 1973

On a clear October day, I bicycle from my laboratory to the campus greenhouse to
gather some rice seeds that I need for an experiment in genetic engineering. The air is
brisk and permeated with a dung-like odor from the dairies nearby. I get off my bike
and enter the silent, humid greenhouse and am quickly surrounded by the revered
grass that feeds half of the world's people (figure 4.1). Rice is grown in more than
89 countries on six of the seven continents. Where rice is the main item of the diet, it
is frequently the basic ingredient of every meal.

I see that the terminal branches of the rice plants are nodding, pulled down
by the weight of the ripening, ovoid grain. The grains are emerald green and still
immature—an optimum time for experimentation. Three to four months from today,
if the experiment is successful, I will have engineered rice plants that are resistant to a
serious disease, called bacterial blight.

I gently harvest a grain-bearing rice stalk with a small, sharp metal knife. Mine
is quite unlike the traditional harvest knives in Southeast Asia. There, knives with
delicate handles carved into a myriad of fanciful forms, such as the Borneo dragon or
boar, are used. Such a knife does "not hurt the rice and therefore is not offensive to the
spirits of the rice or to the Rice Goddess" (Hamilton 2003). Although my simple tool
will not impress the spirits, I am respectful so as to not displease them.

I place the rice stalk carrying the grains in my bag, leave the greenhouse, pick up
my bicycle, and pedal away. On my way back to the lab, I recall the day in graduate
school when I decided to devote the rest of my career to discovering the basic workings
of this plant. I realize that even seventeen years later, the rice plant still enchants and
intrigues me. A few minutes later, I arrive and lock up my bike in front of the lab.
I walk up the stairs, enter the laboratory and settle down in a chair next to the sterile,
white table. I carefully scrape off the green, young hull to expose the immature and
still doughy seed, which carries the precious embryo and the genetic material within.
If left to mature, the embryo would become edible grain.

A quick dip in ethanol, followed by a soak in bleach and a spray of sterile water
will protect the rice embryo from contaminants in the air—bacteria and fungi that are

FIGURE 4.1 Rice Plant with Grain.

harmless to humans but can kill the delicate rice cells. With tweezers, I carefully place about twenty hulled grains onto a freshly prepared plate carrying nutrients that will nourish the embryo. I then move the plates into a growth chamber that will supply adequate light and heat. I have just completed the first step in the process of genetic engineering.

For thousands of years, farmers have deliberately selected and improved plants with desired characteristics from wild and cultivated plants. For example, 10,000 years ago farmers in ancient Mesopotamia developed a hybrid between wild species of wheat and cultivated wheat that became the ancestor of our modern bread wheat (Dubcovsky and Dvorak, 2007). Today breeders manipulate plant species to create desired combinations of traits for specific purposes. In rice and corn, this artificial selection process is generally carried out using pollination. In this process the breeder transfers the male pollen grains to the female part of the flower (box 4.1). Using these techniques, breeders have been highly successful in developing new varieties; so successful in fact that corn varieties only faintly resemble their predecessors and survive only in human-made environments (figure 4.2). As with breeding, the goal of genetic engineering is to alter the genetic makeup of the crop.

BOX 4.1 Plant Breeding, Domestication, and Pollination

Plant breeding: the purposeful manipulation of plant species in order to create desired genetic modifications for specific purposes. This manipulation usually involves either cross- or self-pollination, followed by artificial selection of progeny.

(continued)

BOX 4.1 *Continued*

Domestication of plants: an artificial selection process carried out by humans to produce plants that have fewer undesirable traits of wild plants, and which renders them more dependent on artificial (usually enhanced) environments for their continued existence. The practice is estimated to date back 9,000–11,000 years. Today, all of our principal food crops come from domesticated varieties. Over the millennia many domesticated species have become utterly unlike their natural ancestors. Corn cobs are now dozens of times the size of their wild ancestors, producing seeds that are more digestible (figure 4.2).

Pollination: an important step in the reproduction of seed plants—the transfer of pollen grains (male gametes) from the stamens (male organs) to the plant carpel, the structure that contains the ovule (female gamete). The receptive part of the carpel is called a stigma in the flowers of angiosperms (flowering plants).

Self-pollination: pollen is delivered from the male stamen to the female stigma of the same plant.

Cross-pollination: pollen is delivered to a flower of a different individual of the same species by various of pollinators (e.g., wind, insects or humans). Plants adapted to outcross or cross-pollinize may have prominent stamens to better spread pollen to other flowers (e.g., corn).

Open pollination: pollen is delivered to another plant from a genetically related population. The seeds of open-pollinated plants will grow into plants that are similar to the parents. This is in contrast with hybrid plants. Seed from hybrids segregate for various traits and are not identical to the hybrid parent.

For years, my laboratory has used genetic engineering as a tool to identify genes that control resistance to diseases. When I started this work in 1990, scientists knew the function of only a handful of genes. The science of genetics has progressed with remarkable speed since that time. For instance, in 2004, scientists completed the sequencing of the rice genome by identifying the order of every chemical unit that makes up the genetic information for the cell, an accomplishment achieved for only one other plant species, a tiny plant related to mustard called Arabidopsis (see box 1.1). Detailed computer analysis of the sequence suggests that rice has about 42,000 genes, although we still do not know what most of them do. One of the goals of my experiment today is to figure out the function of one of these genes. Its sequence suggests

FIGURE 4.2 Ancient Ancestor of Modern Corn. The corn that Columbus received was created by the Native Americans some 6,000 years ago by domestication of a wild plant called teosinte. Top: An ancient ancestor of modern-day corn, called Tripsacum (Teosinte). It comes from a plant that doesn't look anything like a modern corn plant. Its seeds are born on the end of one of its stalks, instead of on the body of the plant. Tripsacum produces 10 or 20 seeds per plant. A hammer is needed to break the seed coat to expose the nutritious kernel; something that most of our stomachs are not equipped to do. Bottom: An ear of modern corn. Modern hybrid corn produces several ears each bearing in excess of 1000 kernels. The major difference between these two plants is that modern corn is much more productive and more nutritious. If humans had to depend on the wild relative alone, hundreds, if not thousands, of times more plants would be needed. That in turn would take hundreds or thousands of times more acres, in order to get the same yield. (University of California Division of Agricultural and Natural Resources Statewide Biotechnology Workgroup, 2007)

FIGURE 4.3 Tree with Crown Gall Tumor. (Photo courtesy of Rebecca McSorley)

that it is involved in keeping the plant free from disease, but we can't know for sure until we have tested it in a rice plant lacking the gene.

Two weeks later the grain has grown into a glistening mass called a "callus," a sort of stem cell open to direction, not yet having decided whether to develop into a particular organ, such as a leaf or root. I separate the new callus from the grain with tweezers.

The next step is to introduce a new gene into the cells of the rice callus; to do this I rely on a soil bacterium called *Agrobacterium*. This bacterium can do something that virtually no other organisms can do; it can form a bridge to the plant cell and then transfer some of its own genes across the plant cell wall, then across the membrane and into the nucleus. This ancient process, known to biologists for a century, was only understood in detail over the last thirty years. It is now known that this gene transfer "transforms" the plants into food production units for the bacteria. You can tell if a plant is infected by the appearance of large tumors at the base of the plant. Sometimes the results are dramatic—one of the oak trees on the UC Davis campus has a crown gall tumor the size of a small car (figure 4.3).

Biologists, faithful to their long tradition of manipulation and exploitation, have cleverly removed the bacterial genes that cause the tumor so that the bacteria can infect without disrupting plant growth. They have also figured out that it is possible to replace some of the bacterial genes with genes from other species. The bacteria, unaware that the genes have been swapped, will deliver the new genes into the plant. To carry out this subterfuge, biologists employ other tools of our trade: restriction enzymes that act like tiny scissors to cut out the bacterial genes and ligases that act like glue to insert the new genes into the genome of *Agrobacterium* (figure 4.4).

RECIPE 4.1

Isolation of DNA from Organically Grown Strawberries

My friend, Claire Mazow-Gelfman modified a procedure developed by Juniata College to extract DNA from organic strawberries. It is so easy that even her daughter Tamara and her third grade classmates had fun and were successful.

One of the reasons strawberries work so well is that they are soft and easy to mash up. Also, ripe strawberries produce enzymes (pectinases and cellulases) that aid in breaking down the cell walls. Most interestingly, strawberries have enormous genomes. They are octoploid, which means that each cell has eight of each type of chromosome and abundant DNA (Juniata College 2004).

The detergent in the shampoo helps to dissolve the phospholipid bilayers of the cell membrane and organelles. The salt helps keep the proteins in the extract layer so they aren't precipitated with the DNA.

DNA is not soluble in ethanol. When molecules are soluble, they are dispersed in the solution and are therefore not visible. When molecules are insoluble, they clump together and become visible. The colder the ethanol, the less soluble the DNA will be in it yielding more visible "clumping." This is why it is important for the ethanol to be kept in a freezer or ice bath.

Materials (per student group)

Heavy-duty Ziploc bag
1 Organically-grown fresh strawberry
10 ml DNA extraction buffer (soapy, salty water—recipe follows)

Ice-cold ethanol
50-ml Falcon tube
Tooth picks
DNA extraction buffer recipe

- 100 ml Clarifying shampoo mixed with 100 ml water
- Add a pinch of salt to the water/shampoo mixture

Note: The ethanol must be at least 90% and it needs to be cold. Using a plastic pipette makes it easy to dispense. Cut squares of cheesecloth (two layers thick) large enough to hang over the edge of the Falcon tube. This activity can be completed in one 40-minute class period.

Directions

1. If the green leaves on the strawberry have not yet been removed, do so by pulling them off.
2. Put the strawberry into the Ziploc bag and mash for about 2 minutes. You need to completely crush the strawberry. You do not want this mixture to be really bubbly. The fewer bubbles the better.
3. When you're finished mashing, put 10 ml of the DNA extraction liquid into the bag.
4. Mash for another minute. Be careful not to make too many soap bubbles.
5. When you're finished, place the cheesecloth over the Falcon tube.
6. Open the bag and pour some of the mixture through the cheesecloth and allow it to filter into the test tube. Allow only about 3 ml of liquid to filter through into the test tube.
7. Next, carefully and slowly pour ethanol into the test tube filling it to 8.5 ml.
8. Watch for the development of several large air bubbles that have a white cloudy substance attached to them. The cloudy substance is DNA.
9. Take a toothpick and spin and stir it like you're making cotton candy. If you tilt the test tube, you'll get more DNA.
10. Pull out the DNA. It will look like mucus or egg white. As it dries, it will look like a spider web. The fibers are millions of DNA strands.

A few weeks earlier, I used this cutting and pasting method to introduce a gene from a wild rice species, known to be resistant to disease, into *Agrobacterium*. I now dip the callus into a broth containing the engineered bacteria. At this point, the bacterium acts like a courier delivering the gene from the wild rice species into the genome of the cultivated rice species. I imagine that I can see the bacteria go into action, first identifying its target, then infecting the cell, transferring its DNA across the rice cell

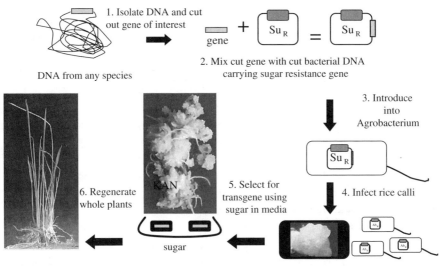

FIGURE 4.4 How to Genetically Engineer a Plant.

1. To isolate DNA, a soapy solution is used to break the cell membranes followed by a salt/ethanol precipitation (see recipe 4.1). The purified DNA is placed in a small plastic test tube and mixed with a set of commercially available "restriction enzymes" that are chosen for their ability to recognize a particular DNA sequence and then precisely cut at that site. By using a combination of these enzymes, it is possible to "cut" a gene out of the genome. The gene fragment can be separated from other fragments of DNA by separating them on a gel or by amplification using the polymerase chain reaction (see chapter 9, Table 4.1). This cutting or amplification leaves a pair of bases exposed, a sort of sticky end.

2. The sticky ends can be joined to the matching sticky ends of a vector DNA that was cut with the same restriction enzymes using ligase enzymes that "glue" the DNA fragments together. The vector carries a Sugar tolerance (Su_R) marker.

3. The genetically engineered vector is then introduced into *Agrobacterium tumefaciens* using an electric pulse (called electroporation). The pulse is thought to disrupt temporarily the membrane and to transmit the vector DNA into the *Agrobacterium* cells. In a small percentage of the cells, the introduced DNA becomes integrated into the host chromosomes or established in the cell as a circular extrachromosomal element.

4. The newly engineered Agrobacterium can be used to "infect" the rice calli.

5. When the infected rice calli are placed in media containing a certain sugar that rice can normally not tolerate, only the infected calli with the new gene and the Su_R marker will begin to grow roots and shoots. The uninfected calli, which do not have the Su_R marker, will die.

6. The surviving genetically engineered cells are regenerated into whole plants.

membrane into the nucleus, and finally into DNA that is bundled into chromosomes in each nucleus. In nature and in the lab, the bacteria do the work of gene delivery.

Scientists cannot predict where the new gene will land beforehand, although it is straightforward to determine the location of the new gene after it is integrated into the crop DNA. Sometimes the gene will land in a spot that disrupts a critical function of the rice cell, in which case that cell will no longer grow. In most cases, however, the transformed cell will thrive and reproduce carrying a new bit of genetic material along with it. How unnatural is this? If we insert genes into random sites in a genome, won't we destabilize a structure that has evolved over millions of years?

It turns out that plant genomes are used to this kind of abuse. It is now well documented that rice and other organisms contain pieces of DNA that move around (called transposable elements) in a seemingly erratic fashion. Not only do they insert themselves into new places, but sometimes they pick up pieces of other genes and take those fragments along for the ride as well. For example, in a recent study it was shown that in the rice genome there are over 3,000 of these pieces of DNA-containing fragments, called pack mules, from more than 1,000 genes (Jiang et al. 2004). Sometimes several fragments are picked up from different genes, rearranged, fused, and then expressed as new proteins. By looking at the genome sequences, we also now know that plants have acquired genes from many different organisms (Brown 2003). It seems then that the genetic engineering process we carry out in the lab has certain similarities to that which occurs in nature.

Not all of the bacteria will be successful in transferring their DNA to the cell; only a very few of the rice cells will receive the new gene and be "transformed." How then does the biologist distinguish the genetically engineered cell from the thousands of rice cells lacking the genes? In fact, this would not be possible without another tool, called the marker gene. In early experiments of plant transformation, the commonly used marker was a gene that encoded resistance to an antibiotic. In this process, not only is the new gene transferred to the cell, but the marker gene is as well. Today, other markers are available, such as those that allow the transformed plant to grow on high levels of particular sugars; in essence this marker is a sugar enablement gene that allows it to grow on sugars it cannot otherwise use (Lucca et al. 2001). By placing the infected rice cells on the "selection" (sugar or antibiotic) media only the transformed cells will survive, inhibiting growth of the rest of the plant cells. In other words, the marker genes bestow properties of survival only to those cells that are genetically engineered, allowing biologists to pluck the newly transformed cells from a lawn of dying, untransformed cells.

Two weeks later I see that the newly transformed cells have thrived despite the high concentration of sugar, and have given rise to new cells carrying the marker gene and the gene of interest. These new cells appear as tiny clumps of whitish, nearly translucent globs about the size of small beads. These cells are genetically identical to the rice seed I began with, except that they also possess a gene from a wild species of rice and the marker gene. I transfer the cells to nutrient plates containing plant hormones that will induce the formation of roots and shoots, give the plants my blessing, and place them back into the growth chamber.

A month later, I arrive early at the lab and hurry over to the growth chamber to check on my tiny plants. Hurrah! The genetically engineered cells have produced shoots and roots, the result I was anticipating. I carefully transplant the GE seedlings into soil-filled pots.

The planting is my favorite step as it recalls my first experiments with clonal propagation at age twelve. I would dig loose, wormy soil from my mother's garden patch; she, like her parents before her, is an avid gardener. I would carry it to my room, spilling soil through the house along the way, and into my closet where I stored my clay pots and a small white plastic jar. The jar contained a plant hormone that would induce rooting. It was called Rootone and is still sold today to household gardeners to induce root growth from cuttings. I would carefully cut the shoot of an African violet and dip it into the Rootone until it was evenly dusted with white powder. Then I buried the stem in the freshly watered soil and placed it on my windowsill. After very careful tending and watering, roots would grow, new shoots would emerge, and I would have a genetically identical plant—a "clone"—to add to my collection of purple flowered plants. Without this capability of plants to regenerate new tissues, such as roots and shoots, plant propagation would not be possible, nor would genetic engineering.

Today, I carry the newly potted GE rice plants to the greenhouse, set them down, water them, and head home. On the way, I reflect that although I no longer work in my closet with toys strewn about, the basic components I worked with as a child are the same: plants, soil, hormones, and genes. The difference is that at the end of this experiment I will have determined if the gene I am studying is the one that makes the wild species resistant to disease, and, more importantly from a practical point of view, I will know if I can engineer resistance in the cultivated species simply by adding one gene from the wild species.

One month later, I head back to the greenhouse with two of my students. After entering the greenhouse, we begin clipping off the tips of hundreds of leaves armed with scissors dipped in the disease-causing bacteria. The scissors cause small wounds that the bacteria use to infect the plant. When we are done, the floor of the greenhouse looks like a busy new-age barbershop, with green locks everywhere.

Ten days later we return and check the plants. The control plants (those lacking the new gene) are very sick, with long watery lesions traveling down the length of the leaves. Those that we genetically engineered are green and healthy. I cannot believe my eyes—I had not expected such dramatic results. I ask my students if perhaps we had made a mistake and clipped the leaves with water instead of the bacteria, but they just smile and shake their heads. Our attempt to genetically engineer rice for resistance to bacterial blight disease is a success (Song et al. 1995). This discovery sheds light on how the rice plant has thrived, and therefore nourished humans, for 8,000 years.

It was in the early 1970s that researchers in the San Francisco Bay Area first demonstrated that it was possible to genetically engineer bacteria with a new trait. They showed that genes from different species could be cut and pasted together and that these new "genes" could be grown and expressed in the bacteria. The famous 1975 Asilomar International Conference on the use of recombinant DNA technologies gathered 150 prominent scientists in the field to debate issues of safety, risk, costs, benefits, and regulation (table 4.1).

As requested by these scientists, the director of the National Institutes of Health (NIH) established an advisory committee to explore the benefits and risks of the new field, develop procedures for minimizing risks, and draft guidelines for research. Recently, Susan Wright, a historian of science at the University of Michigan, suggested that in addition to developing the regulatory framework to move ahead with the research, "the public had to be persuaded that the fruits of genetic engineering would benefit everyone, not just scientists." This dual approach was successful: "[the] new field of research flourished and as a result we now understand a great deal more about the complexity, fluidity, and adaptability of genomes—to an extent that was unthinkable at the time of the 1975 conference" (Wright 2001).

Today the safety and benefits of genetic engineering are being reevaluated, even as applications of genetic engineering proliferate, and have already profoundly changed crop production practices and medicine worldwide. For example, human insulin, the first GE drug marketed, has been used since 1982 for treatment of diabetes, a disease affecting over 7% of the U.S. population (National Diabetes Information Clearinghouse 2005). GE insulin has replaced insulin produced from farm animals because of its lower cost and reduced allergenicity.

Today, millions of people worldwide are treated with GE medicines. One billion acres of GE crops have been grown; hundreds of millions of people have eaten GE food for more than a decade without a single verifiable case of adverse side effects to the environment or to human health. Still, GE provokes controversy, and, sometimes, violent protests (Guyotat 2005).

The debate about genetically engineered crops offers a good example of the concerns that people have about the safety and benefits of powerful and potentially transformative new technologies. Have there been other scientific advances that have provoked so much controversy and presented such a radical alternative to established practices? If so, was the technology ever completely accepted and integrated into everyday life? And finally, was it useful?

Two hundred years ago, many people believed that "natural" or organic compounds (those containing carbon and hydrogen) isolated from plants and animals were fundamentally different from those that were derived from minerals (called inorganic compounds). They thought organic compounds contained a "vital force" that was only found in living systems.

TABLE 4.1 Highlights in the History of Biological Technology

Year	Scientist(s)	Discovery
4000 B.C.E.	The Chinese	Cultivated rice along the Yangtze River
1750 B.C.E.	The Sumerians	Discovered how to brew beer
250 B.C.E.	The Greeks	Practiced crop rotation to maximize soil fertility
1859	Charles Darwin	Published *The Origin of Species*
1866	Gregor Mendel	Demonstrated inheritance of "factors" in pea plants
1870	Jean-Henri Fabre	Discovered insect pheromones
1888	Charles V. Riley	Imported the Australian lady beetle to control cottony cushion scale on citrus
Early 1900s	Sir Albert Howard	Developed principles of organic farming
Early 1900s	Rudolf Steiner	Developed Biodynamics
1901	Shigetane Ishiwatari	Isolated *Bacillus thuringiensis (Bt),* in Japan
1910	Thomas Hunt Morgan	Received the Nobel Laureate in Medicine in 1933 for his discoveries concerning chromosome in heredity
1911	Ernst Berliner	Rediscovered *Bt,* in Germany
1927	Hermann J. Muller	Awarded the 1946 Nobel Prize in Medicine for the studies of mutations induced by X-ray irradiation
1928	Fred Griffith	Proposed that some unknown "principle" had "transformed" the harmless R strain of *Diplococcus* to the virulent S strain.
1944	O. Avery, C, MacLeod Maclyn McCarty	Reported that they had purified the transforming principle in Griffith's experiment and that it was DNA
1947	J. I. Rodale	Established an experimental organic farm in Pennsylvania
Late 1940s	Barbara McClintock	Developed the hypothesis that transposable elements, pieces of DNA that move from one place to another in a genome, can explain color variations in corn. In 1983, was awarded The Nobel Prize in Physiology or Medicine
1951	Rosalind Franklin	Obtained sharp X-ray diffraction photographs of DNA, leading to the discovery of the structure of DNA by Watson, Wilkins, and Crick who were awarded the Nobel Prize in Physiology or Medicine in 1962
1952	M. Chase, A. Hershey	Provided final proof that DNA is the molecule of heredity.

(continued)

TABLE 4.1 *Continued*

Year	Scientist(s)	Discovery
1956	H. F.-J. Angus	Paved the way for using *Bt* toxin as an organic pesticide
1962	Rachel Carson	Writes *Silent Spring*
1967	Alan Chadwick	Started UC Santa Cruz Student Garden
1967	Harry Shorey	Discovered pheromones can be used for mating disruption in cabbage looper moths
1970	Hamilton Smith Kent Wilcox	Isolated the first restriction enzyme that could specifically cut DNA molecules. Smith received the 1978 Nobel Prize in Medicine
1972	Paul Berg	Produced the first recombinant DNA molecules. Berg received the 1980 Nobel Prize in Chemistry for work with DNA (shared with Walter Gilbert and Frederick Sanger for their development of DNA sequencing techniques)
1973	S. Cohen, A. Chang, H. Boyer, R. Helling	Published a paper in the Proceedings of the National Academy of Sciences showing that a recombinant DNA molecule can be maintained and replicated in *Escherichia coli* ("cloning"). Boyer received the 1997 Nobel Prize in Chemistry for uncovering the enzymatic mechanism for synthesis of ATP
1975	International meeting at Asilomar, California	Urged the adoption of guidelines regulating recombinant DNA experimentation
1977	Genentech	Used recombinant DNA methods to make medically important drugs, including human insulin
1980	USDA	USDA Report and Recommendations on Organic Farming
1980	The U.S. Supreme Court (Diamond v. Chakrabarty)	Ruled that genetically altered life forms can be patented
1985	Kary B. Mullis	Published a paper describing the polymerase chain reaction (PCR), the most sensitive assay for DNA yet devised. Together with Michael Smith, won the Nobel Prize in Chemistry in 1993
1987	Advanced Genetic Science	Field-tested "Frostban"—a GE bacterium that inhibits frost formation on crops; the first outdoor tests of a GE organism
1987	National Academy of Science	Concluded that transferring genes between species posed no serious environmental hazard

(continued)

TABLE 4.1 *Continued*

Year	Scientist(s)	Discovery
1987	U.S. Congress	Established LISA (Low Input Sustainable Agriculture) Program (later became the Sustainable Agriculture Research and Education Program)
1988	The Human Genome Project	Established the goal of determining the entire sequence of DNA composing human chromosomes
1989	National Academy of Science National Research Council	Reported on the Role of Alternative Farming Methods in Modern Production Agriculture
1990	U.S. GAO	Reported on Alternative Agriculture: Federal Incentives and Farmers' Opinions
1992	Bob Scowcroft	Established the Organic Farming Research Foundation (OFRF)
1992	United States Department of Agriculture	Approved for commercial production, "FlavrSavr" tomatoes, genetically engineered to have a longer shelf life
2000	Arabidopsis Genome Initiative	Completed sequence of the *Arabidopsis thaliana* genome
2002	International Rice Genome Sequencing Program	Completed detailed genome sequence of rice
2004	The National Academy of Sciences Institute of Medicine	Concluded that biotech crops pose risks similar to other domesticated crops
2005	Farmers	Planted the billionth acre of GE crops
2006	Andrew Fire and Craig Mello	Awarded the Nobel Prize in Physiology or Medicine for their discovery of RNA interference—gene silencing by double-stranded RNA

Sources: Lane 1994, http://www.accessexcellence.org/AE/AEPC/WWC/1994/geneticstln.html; http://www.ucbiotech
.org; http://www.fao.org/DOCREP/006/Y4011E/y4011e0e.htm; http://cuke.hort.ncsu.edu/cucurbit/wehner/741/
hs741hist.html

All this changed in the nineteenth century, when chemists in the United States developed a technique for synthesizing organic compounds from elemental materials. In the first such successful experiment, Friedrich Wöhler (1800–1882) was able to synthesize urea, an organic compound ly produced in an animal's liver.

Wöhler excitedly wrote to Jöns Jakob Berzelius (1779–1848), a Swedish chemist: "I must tell you that I can prepare urea without requiring a kidney of an animal, either man or dog" (McBride 2003). Indeed, his work set in motion a series of experiments demonstrating that compounds synthesized in the laboratory or isolated from nonliving sources were the same. By the end of the nineteenth century, organic synthesis was widely accepted and the vital force theory was abandoned. Wohler predicted this outcome in his letter to Berzelius by saying that he had witnessed "the great tragedy of science, the slaying of a beautiful hypothesis by an ugly fact" (Brooke 1995). Instead of being viewed as the study of substances from living sources, organic chemistry was now seen to be the study of carbon compounds (Bodner Research Web 2004). Rapid advances in organic chemistry quickly led to the synthesis of a variety of organic compounds such as sugars, starch, waxes, and plants oils, as well as drugs, dyes, plastics, pesticides, and superconductors from inorganic materials.

Despite increasing acceptance by scientists, some people viewed the synthesis of natural compounds as "unnatural." Indeed proponents of the vital force theory can still be found on a quick scan of the Internet more than 100 years after it was discredited as a scientific theory. For example, vestiges of the vital force theory linger in the belief that vitamins from natural sources are somehow healthier than vitamins that are synthesized.

By providing us, for the first time, with readily available chemicals, organic synthesis allowed for breakthroughs that we now take for granted in many areas of research and industry, including medicine. For example, the chemical structure of penicillin, determined by Dorothy Crowfoot Hodgkin in the early 1940s, enabled its synthetic mass production. From that point on, penicillin was used routinely to treat bacterial infection and has since become the most widely used antibiotic to date. Beginning with World War II and continuing through today, penicillin has saved an untold number of lives. The discovery that natural penicillin could be further modified chemically led to the expansion of the role of antibiotics in medicine. Modern semi-synthetic penicillins, such as ampicillin or carbenicillin, are now routinely used to treat infection.

Clearly, the synthetic process could be put to good use; however, some products of organic synthesis have caused problems. For example, urea was used as a component of fertilizer and animal feed, providing a relatively cheap source of nitrogen to promote growth. Yet, in the 1950s, the overuse of fertilizers and pesticides began to have significant negative impacts on the environment. This is because excessive fertilizer use can contaminate nearby water sources by promoting algal growth. The extra algae consume more oxygen to live, and their decay deprives other species of oxygen

(Clark 1999). Organic farming evolved partly in response to this overuse, and today, not surprisingly, many organic farmers are skeptical of new technologies that promise to "transform" agriculture.

⸻

Like the advent of genetic engineering, the synthesis of organic compounds is an example of technology with vast potential, which has led to enormous change in our way of life. As with genetic engineering, many viewed the process as "unnatural." Did the benefits of organic chemistry outweigh the risks?

This is not always easy to tease apart. My colleague Dr. Andrew Waterhouse, who is a professor of viticulture at UC Davis, points out a basic problem: "technology advances are so interconnected within our daily lives that it is hard to see how you can subtract one without causing the whole system to stop functioning." For instance, today almost "everything" is dependent on computers. Writing this book requires that I type pages and pages of prose on my computer and then cross out most of it as I continually make revisions. I also rely on frequent communication with my husband and my assistant, Rebecca, through the Internet. In fact, this book would not exist without a whole array of "high performance" plastics, computer chips, and special dyes for computer screens. Without the related chemistry, we could not have computers or virtually any other modern communications. The same goes for developing novel energy technologies. Raoul spent the last few weeks installing solar panels to generate energy for our household. These photovoltaic cells could not have been made before the discoveries of Wöhler and those of subsequent scientists' work. By facilitating the conversion of the sun's energy into electricity, Wöhler has affected our lives in ways that he likely never dreamed of.

⸻

What about genetic engineering of plants? Is GE appropriate to use on our food? What sort of criteria can we use to assess its benefits? An appropriate technology, as asserted by the economist Schumacher in his book *Small is Beautiful,* should promote values such as health, beauty, and permanence (Schumacher 1973). Low cost and low maintenance requirements are also of prime importance in Schumacher's definition. Considering both Schumacher's observations and our goals for ecological farming listed in box P.3, it is apparent that GE will sometimes be appropriate for food modification and sometimes not. This is because GE is simply a tool that can be applied to a multitude of uses, depending on the decisions of policy makers, farmers, and consumers.

Still, as we attempt to show in this book, GE comprises many of the properties advocated by Schumacher. It is a relatively simple technology that scientists in most countries, including many developing countries, have perfected. The product of GE technology, a seed, requires no extra maintenance or additional farming skills. Its arrangement of genes can be passed down from generation to generation and improved along the way. It is therefore clear that humans will likely reap many significant and life-saving benefits from GE. This is because even incremental increases in the nutritional content, disease resistance, yield, or stress tolerance of crops can go a long way to enhancing the health and well-being of farmers and their families. There is also potential for applications of GE to reduce the adverse environmental effects of farming and enable farmers to produce and sell more food locally. Indeed, as described in box 4.2 and discussed in future chapters, there are data to suggest that the use of GE has already drastically reduced the amount of pesticides sprayed worldwide, saved the U.S. papaya industry, and provided new tools to save the lives of impoverished children (Toenniessen 2003).

BOX 4.2 **Genetically Engineered Papaya Saves an Industry**

In the 1950s, the entire papaya production on the Island of Oahu was decimated by papaya ringspot virus, which causes ring spot symptoms on fruits of infected trees. Because there was no way to control the disease, the papaya farms moved to the island of Hawaii where the virus was not yet present. By 1970, the virus was discovered in the town of Hilo, just twenty miles away from the papaya growing area, where 95% of the state's papaya was grown. Because it was very probable that the virus would eventually enter the growing area, in 1978, Dennis Gonsalves, a native of Hawaii, and coworkers initiated research to develop strategies to control the disease. In 1992, the virus was discovered in the papaya orchards and by 1995 the disease was widespread, creating a crisis for Hawaiian papaya farmers. Fortunately, Gonsalves' group was able to develop papayas resistant to the virus by using GE.

Gonsalves' group spliced a small snippet of DNA (made from viral RNA; called RNA interference; see box 12.1) from a mild strain into the papaya genome. Similar to human vaccinations against polio or small pox, this treatment immunized the papaya plant against further infection. The GE plants were resistant to the viral strain, and were crossed to other papaya varieties to generate an abundance of GE seed. In May 1998, the GE seeds were distributed at no cost to local growers. The GE papaya yielded twenty times more papaya than the non-GE variety in the presence of the virus. By September 1999, 90% of the farmers had obtained transgenic seeds, and 76% of them had planted the seeds.

Funded mostly by a grant from the USDA, the project cost about $60,000, a small sum compared to the amount the papaya industry lost between 1997 and 1998, prior to introduction of the transgenic papaya. By 1998 papaya production had dropped to 26 million pounds. After release of GE papaya to farmers, production rapidly increased with a peak of 40 million pounds in 2001.

(continued)

BOX 4.2 *Continued*

Interestingly, in addition to helping conventional papaya growers, the GE papaya has also benefited organic growers. This is because the initial large-scale planting of transgenic papaya together with the elimination of virus-infected fields drastically reduced the amount of available virus inoculum. Also, growers can continue to harvest fruit from non-GE varieties by planting non-GE papaya in the center of a large circle of GE papaya for protection (Gonsalves 1998). These non-GE papayas can be exported to Japan, a major export market, which does not accept the GE variety (National Agricultural Statistics Society 2002). In the United States some consumers still prefer the non-GE and certified organic papaya and will pay more for it. They may not realize, however, that if infected, the non-GE organic papaya contains large amounts of viral RNA and protein as compared to the GE papaya, which is virtually free of the virus (Tripathi et al. 2006).

The story of Hawaiian papayas is an example where GE was the most appropriate technology to address a specific agricultural problem. There was no other technology then to protect the papaya from this devastating disease, nor is there today.

The Swiss Army knife and the molecular scissors are examples along a continuum of new technologies developed through human endeavor and creativity. Which one of these technologies is truly "appropriate" for agriculture? There is no simple answer to this question. When the goal is a productive and ecologically-based farming system, there are usually many interwoven possibilities. As the physicist and philosopher Jacob Bronowski pointed out fifty years ago, "We live in a world which is penetrated through and through by science and which is both whole and real. We cannot turn it into a game by taking sides....No one who has read a page by a good critic or a speculative scientist can ever again think that this barren choice of yes or no is all that the mind offers" (Bronowski 1956).

Consumers

Five

LEGISLATING LUNCH

Perhaps like oil and water, science and politics do not mix. Or so I wonder as I gaze out the window of my friend Beth's Toyota Camry as she steers us along the winding roads of the Sonoma Valley wine country on the way to our annual yoga retreat. The texture of the slightly rolling hills is a refreshing contrast to the flatness of the Central Valley where we both live. The sun shines through gaps in the rain clouds illuminating the startlingly brilliant fall foliage of the vineyards. The pumpkins balanced on the farmers' fence posts look as if they have recently been immersed in a dye extracted from the turning leaves. In view is a lovely Victorian farmhouse set back from the road. In the tidy yard, a sign proclaims: "Yes on Proposition M." If passed, the State of California 2005 initiative, Measure M, "would, for at least the next ten years, prohibit the raising, growing, propagation, cultivation, sale, or distribution of most genetically engineered organisms in Sonoma County" (Sonoma County, California 2005).

I hope that voters know that the beauty here is threatened by a tiny bacterium called *Xylella fastidosa* that causes a disease lethal to the vines. It is transmitted by an insect called the glassy-winged sharpshooter. As the insect sucks the nutritious liquids out of the grape leaf veins, it injects the bacterium, which then multiplies, spreads, and clogs the veins that supply the plant with water. The result is mottled leaves on plants that slowly die over a period of years as their once healthy rootstocks are reduced to mush. During severe epidemics, the infected vineyard will look as if it had been scorched by a fast moving fire (Purcell 2006). At this point, destruction of vines and replanting are the only way to save an infected vineyard. The county's 60,000 acres of wine grapes, with an annual value of more than $300 million, are at risk (Cline 2005). Certain grape varieties, including Barbera, Chardonnay, and Pinot Noir, are highly susceptible to this disease. Because there are no known varieties with resistance to the disease, standard breeding for resistance is not an option. And pesticides, even the most toxic, do little to deter the insect (Purcell 2006). Instead, to control the disease, scientists are now trying to genetically engineer the grape vines using a method similar to that successfully used to protect papaya from papaya ringspot virus (see boxes 4.2 and 12.1). If passed, the proposed ban on GE would prevent future planting of grapes that are resistant to *Xylella*, should they be developed by genetic engineering, the most promising area of research for controlling the disease.

Beth, the long-time manager of the local food co-op that buys from many organic farmers in the area, notices the "Yes on Proposition M" sign and says: "I know that several trade associations that represent organic growers, including the California Certified Organic Farmers, support Measure M. Yet, I find it curious that the Sonoma initiative makes exception for genetically engineered insulin and for cheeses made with genetically engineered rennet (box 5.1). It also allows for the buying, selling, and eating of food with GE ingredients (Sonoma County, California 2005). If there is something inherently risky about GE, then why so many exceptions?"

BOX 5.1 **GE Rennet**

Cheese is made by coagulating milk by the addition of rennet to produce curds. The curds are separated from the liquid whey and then processed and matured to produce a wide variety of cheeses. The active ingredient of rennet is the enzyme chymosin. Until 1990, most rennet was produced from the stomach of slaughtered newly born calves. These days, at a cost one tenth of that before 1990, chymosin is produced by genetically engineered bacteria into which the gene for this enzyme has been inserted, and is used for making cheese in the United States, Europe, and other parts of the world.

I answer, "Farmers that grow GE corn here buy the seed from the Monsanto Corporation. Many consumers have not forgotten that Dow Chemical and Monsanto were the two largest producers of Agent Orange for the U.S. military during the Vietnam War." Although not mentioned in the initiative, the indirect presence of this large multinational corporation in the Sonoma Valley is a major issue. In fact, this may be the underlying reason for the proposed ban. Many consumers fear the power, and question the business practices, of this company.

Beth says, "You are right, they have definitely not forgotten. Few people are willing to trust the maker of Agent Orange to genetically engineer our food."

I ask, "Do you think the reaction to genetic engineering would have been different if it had first come to public attention after it was used to develop crops that benefited poor farmers or malnourished children, or if the first products had been funded by nonprofit agencies, like the GE papaya?"

"Almost surely," Beth says.

Beth and I further contemplate the motives behind the initiative. We know that in addition to a perceived external manipulation of their food by multinational biotech companies, some consumers are also concerned about the safety of the genetic engineering process itself. Others fear that GE crops will economically erode the organic foods niche, because with some GE crops the farmer can apply fewer pesticides—one of the advantages of organic farming. They also are motivated by a desire to isolate organic crops from errant pollen that may stray from genetically engineered crops.

FIGURE 5.1 Sweet Corn Infected with Corn Earworm. On the left are three ears of late-season organically grown sweet corn. On the right are three ears of GE sweet corn containing Bt. (Courtesy of Fred Gould, North Carolina State University)

Because there is some organic corn grown in Sonoma County, there is a possibility of cross-pollination between GE corn and organic corn (Cline 2005). Although an organic grower has never been decertified due to the presence of transgenes in their seed—indeed such decertification would be contrary to the clear standards set by the USDA—(Hawks 2004), some organic business owners worry that consumers will reject their crop if it should test positive for the presence of a transgene. They may not realize that testing is neither required nor encouraged by the USDA National Organic Program.

Some organic growers do not see GE as benefiting their businesses or customers. Yet, organic growers do have pests they need to control, and they could benefit from some genetically engineered crops. For example, in the Central Valley, organic sweet corn does not rank as one of the top twenty organic crops sold, largely due to the corn earworm pest (California Department of Food and Agriculture 2004). An ear of corn usually contains one or two plump, well-fed, yellowish worms; this has prompted growers nationwide to either accept wormy corn or apply broad-spectrum pesticides up to twenty times per crop. The Sustainable Agriculture Research and Education Foundation reports that organic growers are forced to offer one of their most profitable summer crops complete with extra, unwanted protein. "When the earworm hit, sales would drop considerably," said Steve Mong, a vegetable grower in Stow, Massachusetts. "We would leave a knife on the table so anyone who didn't want to take a worm home with them could cut it out" (Sustainable Agricultural Research and Education 2002). The alternatives are pesticide-treated crops or transgenic sweet corn varieties that are free of the corn earworm (figure 5.1; Syngenta 2004).

Despite the stance of the organic trade organizations, some individual organic farmers would like more specifics about genetic engineering before outright rejection of the technology. Our friend Frances Andrews is an organic farmer in our Central Valley community. She earned a history degree from Duke University and worked for Morgan Stanley in New York City before coming west in 1986. She worked at the famed Chez Panisse and Café Fanny restaurants in Berkeley, California, and in 1993, founded a seventy-acre organic farm with her husband. They grow rosemary, lavender, parsley, and other fresh herbs; cherry and heirloom tomatoes; nuts; and a whole array of other fruits and vegetables.

Frances has been casually following the debate on the use of genetic engineering for ten years. The other day she said, "I am getting more confused about genetic engineering as time goes on. I have heard things about genetic engineering that bother me, but then they turn out not to be true. I think people are making conclusions when they don't have the facts. They are trying to make the issue black and white when it is actually gray. It does not need to be one side against another, all good or all bad. And I also hear things that are positive. For example, I have heard that Chinese and Indian farmers growing GE cotton have reduced their use of pesticides dramatically (Bennet et al. 2006; Huang et al. 2005; Huang et al. 2002; Qaim and Zilberman 2003). If this is true, how can I *not* feel like that is a good thing?"

Next to a small barn converted into a winery we notice a gas station. We pull in and get out of the car to stretch. Beth notices a local flyer asking voters to support Measure M. It pictures the destruction in New Orleans wrought by Hurricane Katrina and the bewildered gaze of U.S. President George W. Bush. The flyer proclaims "Who do you trust with your family's health and safety? When FEMA failed, more than a million Americans suffered." It strikes us both that the publicity is off the point, aimed more at frightening consumers than helping voters understand the issues. Of course, the government's response to the flooding of New Orleans has nothing to do with GE foods. Furthermore, I suspect President Bush is uninformed about this local initiative as he is not known for his expertise in science or agriculture. I dislike this kind of fearmongering through the media, as well as the general lack of scientific scrutiny that the flyer reflects.

But as my colleague Sarah Hake, a corn geneticist at UC Berkeley, says, "Fear sells; data do not. Simple successes of genetic engineering in agriculture are not often heard in the popular press—rather, we are given a smorgasbord of reasons to be afraid. Supporting anti-GE measures simply shuts the door rather than allowing important questions to be asked about the environmental and food safety risks and benefits of GE crops" (Bolinas Heresay News 2004).

As we climb back in the car Beth says, "Some of my customers are truly afraid of eating food that contains even minute amounts of GE ingredients. Every day one of them asks about the issue. The questions usually do not come from farmers, but from customers quite removed from farming. To many of them, the GE process seems unnatural and does not fit with their concept of traditional farming. Those most closely associated with living from the land, however, the farmers themselves and their families, are increasingly interested in genetics and the possibilities that GE promises, not wholeheartedly embracing the concept, but interested. It makes me wonder if urban and suburban residents want to maintain a sense of mystique about an unchanging rural life so that they have a dream in which to retreat when their lives become too hectic."

Her comments ring true. Based on what we have seen so far in Sonoma County, the rural–urban divide is evident. The Sonoma County Farm Bureau is opposing the initiative. In contrast, urban residents, food processing companies, and wineries support it, hopeful to include "GE-free Sonoma" on their label, as a new way to market their products. It seems that the images of a farmer working the land, the cows quietly chewing their grass, and the ripening fruit ready for the harvest, represents a sort of life that many long for; a life of order and beauty that is free from pests, stress, and new technologies. Although this may be what is desired, it is not the life most farmers lead.

We also see this division elsewhere in California, with agricultural counties opposing additional restrictions on the use of GE, and other counties favoring them. For example, the board of supervisors in Kern County, California, the fourth largest agricultural county in the nation, recently passed a resolution affirming "the right for farmers and ranchers to choose to utilize the widest range of technologies available to produce a safe, healthy, abundant and affordable food supply, and that the safe, federally regulated use of biotechnology is a promising component of progressive agricultural production" (Cline 2005). Similar resolutions were passed by several other counties in the agriculturally-rich San Joaquin Valley of California, including Fresno County, the largest agricultural county in the nation with almost $3 billion in annual agricultural income (Cline 2005).

Only voters in the California counties Marin and Mendocino, which have far fewer agricultural activities, passed anti-GE initiatives similar to Measure M. In 2002, Marin ranked 41 out of California's 58 counties in terms of total value of agricultural products sold. Mendocino sells even fewer products. The anti-GE laws that were subsequently enacted do not affect current crop production practices in those counties because there were virtually no genetically engineered crops grown there in the first place. Sarah and her husband, Don Murch, an organic farmer in Marin County both opposed the Marin Measure, because as Sarah said in a recent interview, "Although Marin County mostly has organic farms; GE crops can be designed with built-in resistance to pests and disease, thereby reducing the use of pesticides or fungicides. This may never apply to our county, but it could make a difference in other counties where extensive pesticides are used" (Point Reyes Light October 21, 2004).

An hour later, Beth and I enter Mendocino County. We greet our friends, and unpack our sleeping bags and yoga mats and the food. In the kitchen, I sauté eggplants I brought from the student farm with chile, garlic, and olive oil (recipe 5.1). We sit down to eat lunch and drink chardonnay.

It turned out that Sonoma's Measure M would soon be defeated. In an article on proposed bans of genetically engineered food (*Sacramento Bee*, November 4, 2004), the journalist Mike Lee described three similar measures in Butte, San Luis Obispo, and Humboldt Counties in 2004 that were also rejected. For many farmers, the issues were twofold: preventing counties from regulating what they can grow, as well as preserving the possibility of using genetically engineered crops to combat diseases like the one in grapes caused by *Xylella*.

RECIPE 5.1

Spicy Eggplant

INGREDIENTS

2 Eggplants, diced into 1/2" cubes
3 Tbsp Olive oil
1 Clove of garlic, smashed and chopped
1/2 tsp Chile flakes

1. Sauté smashed and chopped clove of garlic in the olive oil.
2. Add the chile flakes to the pan.
3. Add the eggplant to the pan, and sauté until the eggplant is very soft and tender.
4. Add salt to taste.

It seems that the initiative was more an act of defiance, a fight against the change that is ever constant in our lives, rather than a specific, constructive proposal to make agriculture in the county more ecologically balanced. What if agriculture and the communities it supports will neither be lost nor ruined by genetic engineering? What if, instead, GE is a tool that can be refined and shared, as grapes can be fermented and made into wine that delights and nourishes those who drink it?

A few weeks after the yoga retreat weekend, my family has gathered for the Christmas holiday at Tahoe. I am in the kitchen dicing Raoul's organically-grown broccoli, while Anne, my sister-in-law, is making cornbread. Anne lives in Marin County and voted in favor of an anti-GE ordinance in the November 2004 election. The ordinance was opposed by the Marin County Farm Bureau, our friends Sarah and Don, as well as the American Society of Plant Biologists, a nonprofit professional association of which I am a member. The ordinance was passed and now the county deems it unlawful to cultivate, propagate, raise, or grow genetically engineered organisms. I certainly won't be growing any of my rice plants there.

It has been raining for ten days, which means no playing in the snow, so we have plenty of time to talk. The ban is on my mind, so I ask Anne why she supports it. In many ways Anne is a typical resident of Marin. She is highly educated, holds a law degree, tries to make food choices that will support ecologically sound farming, is politically progressive, and currently works part-time for a nonprofit organization dedicated to safeguarding the environment in the Lake Tahoe Basin. Anne is concerned about genetic engineering, reads extensively, and is willing to talk about it. She does not fully trust the scientific community to make decisions about food, and believes that genetically engineered crops were deployed too quickly.

"I voted for the ordinance because it will send a message to the large corporations that the onus is on them to prove their products are safe for human consumption and the environment," she tells me.

I point out that the ordinance contained no language concerning the role of corporations, and that it simply bans farmers from growing genetically engineered crops. I mention, too, that the National Academy of Science and the United Kingdom Genetically Modified Science Review (NAS 2004; GM Science Review Panel 2003) have both already indicated that the crops currently on the market are safe to eat.

"Even if they are safe to eat, I don't like the idea that many of the GE crops grown in the United States are sprayed with herbicides," she adds.

She is referring to the GE crops that are engineered with a bacterial protein that makes them tolerant to the herbicide glyphosate. This is the main component of Monsanto's Roundup, (box 5.2). Conventional farmers like these GE crops because they can spray the herbicide on the weeds. The weeds die, and even if the herbicide drifts onto the GE crop, the crop will survive. Because no hoeing or cultivating is needed, herbicide-tolerant GE crops are now widely grown.

I agree that ideally farmers would be able to control weeds in other ways and avoid spraying anything at all. Besides, it is costly to buy herbicides, and is therefore not useful to farmers in less developed nations who cannot afford the price. But I do know that for all farmers weeds are a big problem and cannot be dismissed lightly. For instance, Raoul tells me that weeds are the main reason why organic rice yields are often lower than conventional yields.

I explain this to Anne and then say, "The good thing about glyphosate is that it is known to be nontoxic to mammals and does not accumulate in water or soil, and is therefore preferable to other widely used herbicides, which persist in the environment."

BOX 5.2 **Herbicide-Tolerant (HT) Crops**

The herbicide glyphosate (trade name Roundup) blocks a chloroplast enzyme (called 5-enolpyruvoyl-shikimate-3-phosphate synthetase [EPSPS]) that is required for plant growth. Crop plants genetically engineered for tolerance to glyphosate contain a gene isolated from *Agrobacterium* encoding an EPSPS protein that is resistant to glyphosate.

Although HT crops do not directly benefit organic farmers, who are prohibited from using herbicides, or poor farmers in developing countries, who cannot afford to buy the herbicides, there are clear advantages to conventional growers and to the environment. In most (but not all) cases, herbicide usage per acre has declined since the advent of herbicide-resistant crops (Cornejo and Caswell 2006), and because glyphosate breaks down quickly in the environment, the overall net effect is a reduction in the toxicity of herbicides used. Glyphosate has a very low acute toxicity (to be poisoned, one would need to ingest three-quarters of a cup of Roundup, which makes it less toxic than table salt!) and is not carcinogenic.

Conventional soybean growers used to apply the more toxic herbicide metolachlor to control weeds of soybeans despite the fact that metolachlor is a known groundwater contaminant and is included in a class of herbicides with suspected toxicological problems. Switching from metolachlor to glyphosate in soybean production has had huge environmental benefits not measured in pounds of active ingredient but in environmental impact (Fernandez-Cornejo and McBride 2002). Another agricultural benefit is that herbicide-resistant soybean has helped foster use of low-till and no-till agriculture, which leaves the fertile topsoil intact and protects it from being removed by wind or rain. Also because tractor-tilling is minimized, less fuel is consumed and greenhouse gas emissions are reduced (Farrell et al. 2006).

"But even if the herbicide is nontoxic, I have read that there is a chemical mixed with the herbicide that can harm fish," Anne responds.

In some of its commercial forms, glyphosate is mixed with a compound called a surfactant that allows it to be more effective. Although glyphosate is nontoxic to freshwater fish, Anne is correct—there is evidence that a surfactant called POEA, used in some formulations, is more toxic than glyphosate alone to aquatic species (Durkin 2003).

I persist on a different tack, "Well, if it is the surfactant you object to, wouldn't it have made more sense to simply ban the surfactant or even the herbicide itself?"

Our polite discussion increases in pace and volume. She responds, "It would be a political dead end to ban the herbicide because a lot of people like to use Roundup in their gardens." It seems to me that she is saying that the herbicide on a small scale is acceptable, but for farmers (i.e., on a large scale) it is not. Also, that because it is popular, we cannot ban the herbicide, only the GE plants that are resistant to it.

I am discouraged. If even my clever sister-in-law is lumping so many disparate issues together, if she does not believe that these distinctions are important, then what chance is there that the rest of the world will be able to assess the complex issues involved? I not so diplomatically suggest that she may have only read the campaign materials about the issue and is therefore not fully informed.

She says hotly, "I have read over 50 hours about this issue and am more informed than most." Ashamedly, I realize that of course this is true. After all, before her third child came along, she used to work as a lawyer, and is accustomed to digging deep into subjects that interest her and forming her own opinion about them.

I hate to feel that I have to convince my sister-in-law of this, or of anything for that matter. But this point seems important. If citizens vote, it should be for a specific matter on which they are well-informed, not because of general concerns about a new technology or the perceived overuse of herbicides. Am I being unfair to persist so long on this matter? Why can't I just relax and have faith that it will start snowing soon, enjoy the beauty of the mountains and of her companionship? Maybe it is impossible to reconcile science and politics anyway, for isn't this after all the point of disagreement? But we have been cooped up too long, so I plunge deeper into it.

"But what about the dramatic reductions of pesticide use in China after the farmers started using genetically engineered Bt cotton?" I ask, "Aren't you pleased that GE has helped reduce the use of pesticides?"

We talk about the results in China where the gene specifying the Bt toxin was genetically engineered into cotton, making the plant resistant to serious insect pests such as cotton bollworm that can destroy the crop (box 5.3). Planting of this GE cotton, eliminated the use of 78,000 tons of insecticides in 2001 (Toenniessen et al. 2003), almost equal to the entire amount that is applied annually in the state of California (Department of Pesticide Regulation Pesticide Use Reporting 2004). In the United States, adopters of Bt crops have also reduced their use of pesticides (Fernandez-Cornejo and Caswell 2006).

Anne responds, "I am certainly very concerned about pesticide use, but we can't ban those either, because everyone is used to them now and are familiar with the risks. Again, it just won't fly politically. Besides I don't think the Bt toxin used in GE crops has been adequately tested."

This is a point on which I do not agree. The Environmental Protection Agency has indicated that there are no known human health hazards associated with the use of most Bt toxins. Nor do Bt toxins have any known effect on wildlife such as mammals, birds, and fish. In fact, the EPA has found Bt toxins to be of such low risk that it has exempted them from food residue tolerances, groundwater restrictions, endangered species labeling, and special review requirements—one of the reasons organic farmers

BOX 5.3 **Crops Genetically Engineered for Insect Resistance**

The soil bacterium *Bacillus thuringiensis* (Bt) produces proteins called Bt toxins that can kill important plant pests such as some caterpillars and beetles. Bt toxins cause little or no harm to most non-target organisms including beneficial insects, wildlife, and people. Sprayed formulations of Bt toxins are among the favored insecticides of organic growers. Corn and cotton have been genetically engineered to make Bt toxins. These GE crops called Bt corn and Bt cotton were created by inserting into the plants' genetic material the bacterial genes encoding the Bt toxins. While Bt toxin sprayed on leaves quickly degrades in sunlight and does not reach insects feeding inside plants, Bt crops make Bt toxins internally. Thus, Bt crops are effective against insects that bore into stems, such as the European corn borer, which causes more than $1 billion in damage annually in the United States.

First commercialized in 1996, today, Bt crops are the most commonly grown transgenic crops in the world. In 2004, an estimated 200 million acres of GE crops with Bt and/or herbicide tolerance were cultivated in seventeen countries worldwide, a 20% increase over 2003. U.S. acreage accounts for 59% of this amount followed by Argentina (20%), Canada and Brazil (6% each), and China (5%) (Fernandez-Cornejo and Caswell 2006).

Long before Bt crops were developed, Bt toxins in sprayable formulations were used to control insects. This fact allowed the EPA and FDA to consider twenty years of human exposure in assessing human safety before agreeing to register Bt corn for commercial use. In addition to these data, numerous toxicity and allergenicity tests were conducted on many different kinds of naturally occurring Bt toxins. Based on these tests and the history of Bt use on food crops, it was concluded that Bt corn is as safe as its conventional counterpart and therefore would not adversely effect human and animal health or the environment (Opinion on . . . , EFSA 2004).

Bt corn can improve human and animal health by reducing contamination of food by mycotoxins, which are toxic chemicals produced by fungi (Wu 2006). This is possible because Bt corn reduces insect damage that promotes fungal growth.

In the United States, Mexican-American women living on the Rio Grande border region consume a diet heavy in corn tortillas. Consumption of tortillas made from mycotoxin-contaminated corn increases the risk of neural tube defect pregnancies at a significantly greater rate than American women generally. This is because the mycotoxin interferes with the uptake of folate from maternal cells. The risk of such neural tube defects could be reduced by consuming corn tortillas produced from Bt corn varieties (Kershen 2006).

(continued)

BOX 5.3 *Continued*

Mycotoxins can also cause esophageal and liver cancers in humans and are associated with stunting in children. These problems are especially acute in rural Africa where farmers store a year's supply of corn in wicker cribs that are open to the sun, weather, infestation by beetle and weevil larvae, and fungal contamination (Wu 2006).

In the United States, adoption of certain Bt crop plants by U.S. farmers has resulted in the application of fewer pounds of chemical insecticide, and thereby has provided environmental benefits, but the size of the reduction is dependent on the particular crop. Overall, the USDA Economic Research Service found that insecticide use was 8% lower per planted acre for adopters of Bt corn than for nonadopters (Fernandez-Cornejo and Caswell 2006).

As reported by the USDA ERS, fewer insecticide treatments, lower costs, and less insect damage can lead to a significant profit increase for when pest pressures are high (Fernandez-Cornejo and Caswell 2006). When pest pressures are low, farmers may not be able to make up for the increased cost of the GE seed by increased yields. An analysis of forty-two field experiments indicates that nontarget invertebrates are generally more abundant in BT cotton and Bt maize fields than in nontransgenic fields managed with insecticides In comparison with insecticide-free control fields, certain nontarget taxa are less abundant in Bt fields (Marvier et al. 2007). These results support the USDA conclusion that that Bt crops can reduce environmentally undesirable aspects of agriculture, particularly the nontarget impacts of insecticides.

Chinese and Indian farmers growing GE cotton have reduced their use of pesticides dramatically (Bennet et al. 2006; Huang 2002, 2005; Qaim 2003) and the number of pesticide-related injuries has also dramatically decreased (Huang et al. 2005). Despite these positive aspects, one preliminary study of 481 Chinese farmers in five major cotton-producing provinces suggests that after seven years, populations of other insects (such as mirids) that are not targeted by Bt, and which were previously controlled by spraying broad-spectrum pesticides, have increased (Wang et al. 2006). Nevertheless, studies show that there is an overall net reduction in the use of broad-spectrum pesticides by these farmers (Huang 2007). In Arizona, insects such as the sweet potato whitefly (*Bemisia tabaci*) that are not controlled by Bt in cotton fields can be controlled instead by insect growth regulators, which are considered less harmful than broad-spectrum insecticides (Cattaneo et al. 2006).

One drawback of using any pesticide, whether it is organic, synthetic, or genetically engineered, is that pests will probably evolve resistance to it. Scientists and others have been concerned that widespread use of Bt crops

(continued)

BOX 5.3 *Continued*

would create conditions for insects to quickly evolve resistance to Bt toxins. So far, however, field-evolved pest resistance to Bt crops has not been documented (Tabashnik et al. 2003).

Insects have evolved resistance to Bt toxins in the laboratory, yet only one crop pest, the diamondback moth (*Plutella xylostella*), has evolved resistance to Bt toxins under open field conditions (Tabashnik et al. 2003). But this resistance was not caused by Bt crops, rather it occurred in response to repeated foliar sprays of Bt toxins to control this pest on conventional (non-GE) vegetable crops (Tabashnik 1994). Based partly on the experience with diamondback moth and because Bt crops cause season-long exposure of target insects to Bt toxins, some scientists predicted that pest resistance to Bt crops would occur in a few years.

Contrary to expectations of rapid evolution of pest resistance to Bt crops under worst-case scenarios, a long-term study of resistance to Bt cotton in field populations of pink bollworm (*Pectinophora gossypiella*) in Arizona showed no net increase from 1997 to 2004 in pink bollworm resistance to the toxin in Bt cotton (called Cry1Ac). The lack of field-evolved resistance despite extensive use of Bt cotton and rapid resistance evolution in the laboratory suggest the "refuge strategy"—growing refuges of crops plants that do not make Bt toxins to promote survival of susceptible insects—has helped to delay pink bollworm resistance to Bt cotton (Tabashnik et al. 2005).

---

like to use them. I argue that Bt toxin has been used by organic growers for over fifty years, and that, as far as I can tell, my husband is still healthy. For some reason she is not impressed with my statistical sampling and says so.

"Well, one person spraying it is different than millions of people eating it. What if people start to have allergic reactions?" she asks.

In fact, even with such widespread use of Bt toxin-based sprays in the past fifty years, only two incidents of allergic reaction have been reported to the EPA and these were reactions to the sprays used by organic farmers not to the GE variety. In the first incident, it was concluded that the exposed individual was suffering from a previously diagnosed disease. The second involved a person who had a history of life-threatening food allergies. Upon investigation, it was found that the formulation of the Bt spray also contained carbohydrates and preservatives, which previously had been implicated in food allergies (Glare and O'Callaghan 2000). Because of this result some scientists argue that it would be safer to genetically engineer Bt toxin in the plant rather than spray it as organic farmers now do (Bernstein et al. 1999).

Interestingly, the risks people associate with Bt toxin seem to be connected with how it is presented. In one of my classes, I asked my students which agricultural

BOX 5.4 **Which of These Genetically Engineered Products Would You Accept?**

- Soybeans that make more monounsaturated fatty acids and fewer polyunsaturated and transfat fatty acids, providing healthier sources of vegetable oil
- Rice and corn that express vitamin A and promise to reduce blindness and save lives in many developing countries
- Paint from GE soybeans, eliminating the need for chemical modifications that produce toxic byproducts
- Milk produced by cows fed on corn that contains the Bt toxin gene
- Cheese made with rennet produced by genetically engineered microorganisms instead of being extracted from a calf's stomach
- Mangoes from South America produced in a GE tree that slows ripening. Now mangoes can be shipped to the United States, with more profits to poor farmers
- Locally-grown genetically engineered papaya that are immune to papaya ringspot virus and are cheaper than organic papaya (that carry large amounts of viral RNA and protein)
- Cotton shirts made from GE cotton that are sprayed with fewer pesticides
- Tomato fruit sprayed with a dead bacteria carrying a toxin that kills insects
- Wine from grapes produced by GE vines that are resistant to the glassy winged sharpshooter
- Tofu made from GE soybeans that carry a bacterial gene making them resistant to a benign herbicide
- Tofu made from non-GE soybeans that have been sprayed with more toxic herbicides
- GE peanuts that are less allergenic
- Beef from cows fed GE corn with improved protein content (e.g. high lysine corn)
- Low nicotine cigarettes made from GE tobacco
- The anti-cancer drug Taxol produced from GE corn
- Human insulin made by genetically engineered microorganisms using fermentation
- Human insulin made by genetically engineered plants in the field (half the price as above)

Modified from Chrispeels and Sadava 1994, *Plant, Genes, and Agriculture* products they would avoid (box 5.4). The list included GE foods, as well as "tomato fruit sprayed with dead bacteria carrying a toxin that kills insects." Many of the students concluded that they definitely would not eat this product. "It sounds awful," one student said. She was somewhat sheepish to later learn that such Bt toxin sprays are commonly used by organic growers and that they are widely considered to be safe.

I hope to convince Anne that there are some potential benefits of GE, so I decide to fire away with my most powerful ammunition. "There are some pests that cannot be consistently controlled using organic methods. Most ears of organically grown sweet corn carry fat cannibalistic worms and their frass. Wouldn't you rather eat GE corn carrying trace amounts of Bt toxin than eat corn carrying such surprises?"

"In Marin, I pay someone to bake my bread; I can certainly ask the store to remove the worm so I don't have to do it." Anne replies.

I have to concede that she has a point there. Clearly, different communities have different preferences and can afford different amounts of food processing. The Amish people of Pennsylvania, known for their use of simple, appropriate technology and organic farming methods, have chosen to grow GE tobacco because they are able to sell it for a high price and the community harvest supports their way of life (box 5.5). In Marin, the wealthiest county in the state and in the nation, consumers prefer organic food without the worms and are willing to pay someone else to cut them off. But should laws be imposed to regulate such diverse preferences?

BOX 5.5 **Amish GE Tobacco Growers**

The Amish people of Pennsylvania have started to carefully evaluate the usefulness of genetically engineered crops. One report indicates that more than 600 Amish families in Pennsylvania signed up to cultivate 3,800 acres of transgenic tobacco with reduced nicotine content—enough to produce 345 million cigarettes. The influx of cash has been a boon to the community (Davis 2003). Instead of $400 per acre growing corn they can now earn $3,500 per acre growing the GE tobacco. This high-value crop directly benefits the community because it provides local on-farm work for the families, and it might also benefit some consumers if the reduced-nicotine content helps them quit smoking.

I notice that the yellow cornmeal Anne is using is "enriched and degermed" so I try yet another approach. "Anne, not only is that cornmeal highly processed, but it has synthetic chemicals added for nutritional enrichment. It also likely contains GE ingredients; after all, 70% of processed foods have at least one ingredient from GE corn or soybean (CDFA 2003). None of that is natural. How can you feel comfortable using it?"

She looks momentarily startled. "Well I know I will not drop dead tomorrow; I am not that fearful. I trust that the regulators won't let it completely kill us" (box 5.6).

My brother Rick wanders in to check on the chili he is making and says, "It doesn't matter what the scientists think. If the risks and benefits of GE are not explained

---

BOX 5.6 **Regulatory Oversight of GE Crops**

Before commercial introduction, genetically engineered crops must conform to standards set by State and Federal statutes. Under the Coordinated Framework for the Regulation of Biotechnology, federal oversight is shared by the U.S. Department of Agriculture (USDA), the U.S. Environmental Protection Agency (EPA), and the U.S. Food and Drug Administration (FDA).

USDA's Animal and Plant Health Inspection Service (APHIS) plays a central role in regulating field-testing of agricultural biotechnology products. Through either a notification or permit procedure, such products, which include genetically engineered plants, microorganisms, and invertebrates, are considered "regulated articles." APHIS determines whether to authorize the test, based on whether the release will pose a risk to agriculture or the environment. After years of field tests, an applicant may petition APHIS for a determination of nonregulated status in order to facilitate commercialization of the product. If after extensive review, APHIS determines that the unconfined release does not pose a significant risk to agriculture or the environment, the organism is "deregulated." At this point, the organism is no longer considered a regulated article and can be moved and planted without APHIS authorization.

If a plant is engineered to produce a substance that "prevents, destroys, repels, or mitigates a pest," it is considered a pesticide and is subject to regulation by EPA. FDA regulates all food applications of crops, including those crops that are developed through the use of biotechnology, to ensure that foods derived from new plant varieties are safe to eat.

Though the current regulatory system is considered to be effective, USDA, EPA, and FDA continuously look forward and make necessary changes to address new trends and issues of the future. For example, USDA's APHIS has made updates in 1993 and 1997 and is currently considering the need for additional changes in the regulations. The National Academy of Sciences also issued a report suggesting that regulation could be improved by making the process more transparent and rigorous, by enhanced scientific peer review, by soliciting public input, and by more explicit presentation of data, methods, analyses, and interpretations.

Fernandez-Cornejo and Caswell (2006)

clearly to the general public, no one is going to embrace it until they are convinced that the food won't hurt their children or the environment."

Rick lifts the lid on the pot of steamed broccoli, which I have completely forgotten about and which now resembles a green paste. "Are you planning to cook this until tomorrow?" he asks.

We all laugh as I hurry over to turn off the heat. A little tension in the room is released as if I had opened a window to let the steam out. Anne has not really conceded anything, for that is not her style, nor is it often mine. We are known as the stubborn ones in the family. But we do realize that we share similar views on the importance of food safety and reducing the use of pesticides, and in all likelihood have more in common than not. Additionally, based on her willingness to use GE cornmeal in the bread, it seems that we also we do agree that the genetically engineered corn now on the market is safe to consume.

RECIPE 5.2

Cornbread

INGREDIENTS

2 Tbsp Butter

2 Eggs

1/4 c GE canola or corn oil

2 Tbsp Honey (Note: most honey on shelf is from Canadian canola fields, which are 80% GE)

1 c Buttermilk

1 c GE cornmeal (freshly ground is preferable, but not necessary)

1/2 c Whole wheat flour (freshly ground is preferable, but not necessary)

1/2 c Barley flour (freshly ground is preferable, but not necessary)

1/2 tsp Salt

2 tsp Baking powder

1. Preheat oven to 425°F.
2. Put butter into an 8-inch-square pan and set in oven while preheating.
3. Beat eggs together.
4. Add oil, honey, and buttermilk to egg mix.
5. Gently mix in dry ingredients.
6. Quickly pour into pan and bake for 25 minutes.

Adapted from Madison (1997)

The kids are hungry so I take the cornbread out of the oven a little bit early. Beneath the smooth yellow surface all the contradictions of science, agriculture, and politics seem to be hidden. I am surprised that it looks so plain. I dab on a bit of butter, which the steam melts quickly. We each bite into the crumbly yellow bread and all agree it is delicious.

Six

WHO CAN WE TRUST?

Despite my growing belief that the California initiative process (where citizens can pose new legislation directly to voters) is not a constructive means for debate of scientific issues, I cannot dismiss the general anxiety about genetic engineering, and the distrust of science that is reflected by these campaigns. It seems that to successfully make decisions on how to use GE for the betterment of humankind and the environment, the public will need to understand the scientific process and learn to distinguish high-quality scientific research that has stood the test of time and can largely be relied on from simple assertions or unsubstantiated rumors.

Jim Holt, a writer for the *New York Times Magazine*, cites a survey indicating that less than 10% of adult Americans possess basic scientific literacy. For nonscientists, it may be the sheer difficulty of science, its remoteness from their daily activities, "that make it seem alien and dangerous" (Holt 2005). Yet, the societal values that science promotes—free inquiry, free thought, free speech, transparency, tolerance, and the willingness to arbitrate disputes on the basis of evidence—are exactly the qualities needed when debating the future use of GE in generating new plant varieties. In the words of Ismail Serageldin, Director of the Library of Alexandria and past Vice President for Environmentally and Socially Sustainable Development of the World Bank, an understanding of the scientific process is important "not just to promote the pursuit of science, but to yield a more tolerant society that adapts to change and embraces the new" (Serageldin 2002).

Misrepresentation of science for ideological or political purposes simply muddies the debate, and sadly, with respect to the GE foods, this often occurs. For example, to suggest that genetic engineering is dangerous, proponents of the California initiatives to ban the process often cite a book called *Seeds of Deception* (Smith 2004), written by a former Iowa political candidate for the Natural Law Party with no scientific training. This book is the likely source for information on another Sonoma county flyer suggesting that "Lab animals fed GE food develop stomach lesions," in reference to a fundamentally flawed experiment carried out in 1999 that was never confirmed (Ewen and Pusztai 1999). To lend credence to those irreproducible results, Smith cites the experiment of a seventeen-year-old student who fed mice genetically engineered potatoes. According to the referenced Web site, "[the mice] fed GM ate more, probably because they were slightly heavier on average to begin with, but they gained less weight." In addition, "marked behavioral differences" were observed though the boy admitted, "these were 'subjective' and not quantitative." Smith argues that this experiment demonstrates that GE food may have negative effects on the "human psyche" and concludes that the boy "has put the scientists to shame." The implication is that the public can trust this experiment carried out by a student, unhampered by scientific training but not those of the scientific community who pointed out the flaws in the original experiment. Smith ignores the fact that this experiment conducted by a teenager was not subjected to the rigorous methods that are inherent to the scientific process.

So how can the public distinguish rumors from high quality science, determine what an established scientific "fact" is, and what is still unknown? Here are some useful criteria:

1. *Examine the primary source of information.* Is there a reference to the source of information? If not, it cannot be verified. If so, is the source reputable? In the case of the boy and the mice, I found that the reference given for the boy's work was to another Web site, and that that web site referred to even another Web site (Ho 2002). It turned out that the only documentation of this "experiment" was a chance meeting with the boy's mother, who was the source of the "scientific information." "Mum Guusje is very proud of her son. . . ." I wonder why someone would cite a conversation with a boy's mother as a good scientific reference? Either the authors of the book and the Web site lack a basic understanding of science and cannot assess the accuracy of the work, or they simply do not care, or both. But they should care; for this kind of deception only confuses and frightens people. And laws are being passed based on this kind of silliness.

2. *Ask if the work was published in a peer-reviewed journal.* Peer review is the standard process for scientific publications. Peer-reviewed manuscripts have been read by several scholars in the same field (called peers), and these peers have indicated that the experiments and conclusions meets the standards of their discipline and are suitable for publication. In the absence of peer-review the significance and quality of the data cannot be assessed. With no peer-reviewed, published record of the boy's subjective experiment, it is doubtful that normal standard scientific methods were applied.

3. *Check if the journal has a good reputation for scientific research.* If a peer-reviewed paper is cited, where was it published? Is the journal widely respected? One tool that is commonly used for ranking, evaluating, categorizing, and comparing journals is the frequency with which the "average article" in a journal has been cited in a particular year or period. The frequency of citation reflects acknowledgment of importance by the scientific community. High-impact

and widely respected journals include *Science* and *Nature*. Therefore, a citation in *Science* generally suggests scholarly acceptance, whereas publication in a nonscientific or little-known journal does not.

4. *Determine if there is an independent confirmation by another published study.* Even if a study is peer-reviewed and published in a reputable journal, independent assessment is critical to confirm or extend the findings. Even the best journals or scientists will occasionally make mistakes and publish papers that are later retracted. Sometimes there may be outright fabrication that is overlooked by the reviewers and not detected until later (Kennedy 2006). In other cases, the scientific report may be accurate but its significance may be misrepresented by the media. A good example is that of genetically engineered corn and the monarch butterfly controversy that erupted in 1999. A Cornell entomologist, John Losey, published a short paper in the scientific journal *Nature* reporting that monarch butterfly larvae died after eating milkweed plants dusted with pollen from GE corn (Losey et al. 1999). The paper generated intense national and international news coverage transforming the monarch butterfly overnight into a dramatic symbol of what some consumers saw as the dangers of agricultural biotechnology. Subsequent scientific studies, including field trials, showed that the exposure of monarchs to GE corn is fairly small and that the threat to monarchs pales in comparison to risks presented by conventional pesticides (Pew Initiative on Food and Biotechnology 2002). Such misrepresentations or errors are usually discovered by other researchers because most reports, especially if it is exciting news such as a suggestion that genetic engineering kills monarch butterflies or makes mice sick, will be rapidly retested by other scientists. If the data are challenged, the first author then has the opportunity to write another paper refuting the challenge. Although it is a slow process to establish a scientific "truth," a particular scientific conclusion will eventually either gain broad acceptance or be discarded.

5. *Assess whether a potential conflict of interest exists.* Most people would agree that a mother usually believes the best about her son, and that pesky details such as lack of scientific training may not bother her. Therefore, a mother's recommendation represents a clear conflict of interest in such a case. Studies tainted by such undisclosed conflicts of interests are a major concern in the debate about genetic engineering. If governmental regulators were to rely solely on data supplied by parties whose primary concern is not the public good but private interest, then the public would have reason to question the integrity of the research. Similarly, if a person with a strong stance on the use of GE in agriculture is an employee of a for-profit biotechnology or organic industry, such employment should be disclosed because a conflict of interest may exist. (Full disclosure: neither Raoul nor I presently have financial relationships with for-profit food biotechnology or organic industries. Transparency is a wonderful disinfectant when honesty is needed.)

6. *Assess the quality of institution or panel.* Does the report emanate from a University accredited by the U.S. Department of Education or equivalent society? Such information is generally more reliable than that issued from a single individual putting information out on the web. In the United States, government research arms such as the National Science Foundation and the National Institute of Health and professional scientific societies generally provide up-to-date, high-quality information. For example, the American Society of Plant Biologists is a nonprofit professional society devoted to the advancement of the plant sciences. It publishes two world-class journals and organizes conferences and other activities that are key to the advancement of the science. The National Academy of Sciences (NAS) is "an honorific society of distinguished scholars engaged in scientific and engineering research, dedicated to the furtherance of science and technology and to their use for the general welfare." (NAS 2006). Election to the Academy is considered one of the highest honors that can be accorded a U.S. scientist or engineer. These types of nonprofit organizations provide a public service by working outside the framework of government to ensure independent advice on matters of science, technology, and medicine.

7. *Examine the reputation of the author.* Do the author(s) have training in science? If so, have they had formal training leading to an advanced degree such as a Master's degree or doctorate, and have they published widely in reputable journals? If not, then are they working with a reputable scientist(s) to evaluate the data? In the case of the boy and the mice, a university affiliation is hinted at, but it seems that the "experiment" was carried out at home and reviewed primarily by his mother.

You, the reader, are now ready to delve into issues surrounding genetic engineering. Applying these tips about the scientific process, you can now more easily assess the accuracy of media reports. Checking scientific sources can be time consuming, but it is worth the effort because such sources will get you closer to accurate facts about GE than rumor or unconfirmed reports.

Seven

⟫⟫⟫⟫✕⟪⟪⟪⟪

IS GE FOOD RISKY TO EAT?

The risks that hurt people and the risks that upset people are almost completely unconnected.

> PETER M. SANDMAN, risk communications consultant, as
> quoted in the *New York Times* article by Henry Fountain

"Fingernails dug into Yosemite rock, rope pulling painfully at my harness, legs unsteady as a sewing machine, arms trembling. . . . The rope slacks and I know it—I'm going to fall to my death." Amie reads.

In December, members of our writing group are on the couch in a corner of Amie's living room drinking vanilla hazelnut tea with soymilk listening to her story. A large wooden table laden with mandarin oranges, persimmons, and a rice-celery salad with purple tomatoes occupies the other half of the cozy room. In front of us is a coffee table piled high with cheeses and homemade treats including Christollen, a buttery egg bread flavored with citrus, cinnamon, and vanilla—a traditional Ronald family holiday favorite (recipe 7.1).

RECIPE 7.1
⟫⟫⟫⟫✕⟪⟪⟪⟪

Trish's Christollen

INGREDIENTS

2 c Milk
1 c Sugar
2 tsp Salt
1 1/3 c Butter
2 Envelopes of yeast
2 c Flour
4 Beaten eggs
8 c Flour
1 1/2 c Chopped blanched almonds
1 1/2 c Raisins, softened in warm water and drained

1/2 c Currants
Grated rind of 1 orange
Grated rind of 1 lemon
2 tsp Vanilla

1. Scald milk and add sugar, salt, butter. Stir and let cool.
2. Add yeast dissolved in a little water and eggs.
3. Stir in 2 cups of flour. Let rest until bubbly.
4. Stir in eggs and rest of flour until light, but not sticky.
5. On a lightly floured surface, knead in almonds, raisins, currants, rinds, and vanilla until smooth and elastic.
6. Cover dough and let rise until doubled in size; punch down and divide into three parts. Let rest for 10 minutes.
7. Flatten each portion of dough into a 3/4 inch oval thick. Brush with melted butter. Sprinkle with sugar and cinnamon.
8. Fold ovals almost in half. Pinch ends firmly together. Place on oiled cookie sheets. 10. Brush with melted butter and let rise for 1 hour or until doubled in size.
9. Preheat oven to 425°F. Bake at 425°F for 10 minutes.
10. Reduce to 350°F and bake for 40 minutes. Allow to cool.
11. Glaze with a mixture of confectioner's sugar and lemon juice (mixed to the consistency of a thin paste).
12. Decorate with fruit and nuts.

Before we met Amie, she was a Zen priest at the famed Green Gulch Farm in Marin County, a Buddhist practice center offering training in meditation and organic farming. After ten years as a priest, she studied midwifery at the University of California, San Francisco, while single-handedly raising her daughter. She is now a certified nurse midwife in Davis. She delivered our three children and therefore retains near-divine status in our eyes.

One would think that a woman with such a history and reputation would be fearless. But apparently, this is not the case. Amie tells us that she has always been afraid of heights, and had chosen rock climbing to face her fears. The hazard of the sport is apparent. After all, a long fall can kill. But in a twist to Amie's story, it turns out that she was only three feet off the ground and attached by a rope to a skilled partner who loved her. Therefore, the probability of a negative consequence was extremely low. It is clear then that our own sense of risk can frighten us even if the activity is not likely to cause harm.

In an interview published in the *New York Times* in 2006, Peter M. Sandman, a risk communications consultant, said, "The likelihood of being affected by a possible hazard is not the major factor influencing whether a person feels 'outrage.' Instead, factors like control and familiarity are much stronger influences" (Fountain 2006).

Take pesticides for example. Many consumers have grown accustomed to pesticides despite the fact that certain kinds, such as the "carbamate" type, are estimated to poison tens of thousands of people each year, mostly farm workers. Despite the fact that some of these poisonings are fatal, relatively few people worry about pesticides or protest against their use on farms. In a similar vein, the rising prevalence of obesity in children has been linked, in part, to overconsumption of highly sweetened drinks (Ludwig et al. 2001). If trends in obesity continue, one in three American children born in 2000 will go on to develop diabetes in their lifetimes. A few schools around the nation have begun to ban or limit the size of soft drinks available to their students, but it is by no means commonplace. In contrast, just the mention of genetic engineering, a process that has been used for thirty years and so far has not harmed a single person or animal, can cause alarm. The apocalyptic quality of the anti-GE advocacy seems wildly disproportionate to the potential risk, particularly in the context of the benefits.

Unlike fluoride or some types of synthetic or organic pesticides such as rotenone, which are unquestionably lethal to animals at high concentrations, GE traits are composed of the same chemical building blocks (DNA and proteins) that we eat every day. Indeed, these are the same components that Buddha ate 2,500 years ago, and they are what we will be eating 2,500 years from now. That is, if humans survive the increasing overcrowding of our planet, inadequate nutrition, disease, poverty, and pollution of our environment. Within one hour, 98% of the DNA in foods is digested completely (Schubert et al. 1997) and most proteins are digested even faster. In other words, the fluoridated toothpaste on your toothbrush or the soft drinks in your refrigerator likely present greater risks to your health than the genetically engineered papaya you had for breakfast.

It seems we may be worrying about the wrong hazards.

When Amie is finished with her short story, I read an excerpt about anti-GE legislation from this book. The group listens politely until it is time for discussion. Amie skips the constructive critique and moves straight to the point. "How can you be sure that it isn't risky to eat GE foods?"

Of all the concerns about GE food, this question of risk presents the biggest worry for the most people. The U.S. National Academy of Science (NAS) addressed a broader question by asking, "Is the *process* of adding genes to our food by genetic engineering any more risky than adding genes by traditional breeding?"

The answer is no. Virtually everything we eat has been genetically modified in some way, and virtually every food we eat poses some kind of risk, albeit a very, very small risk. The NAS committee determined that both the process of genetic engineering and traditional breeding pose similar risks of unintended consequences (NAS 2004).

When I explain this, Amie wants more detail. "What exactly does the phrase 'unintended consequences' actually mean?"

I go on to explain how, in my lab, we typically take a gene from one rice variety and put it into another rice variety. The committee estimates that the risk of unintended consequences resulting from this kind of work is similar to the risk that results from conventional breeding with two existing rice varieties. On the other hand, transferring a gene from a distantly related species, for example, putting a bacterial or fish gene into a plant, is more risky. That sounds frightening until you realize that traditional methods such as mutation breeding pose even greater risks.

"What is mutation breeding?" asks Matt. "Have I eaten a mutant?"

"In all likelihood, you have," I answer. "With mutation breeding, seeds are put in a highly carcinogenic solution or treated with radiation to induce random changes in the DNA. After germination, surviving seedlings that have new and useful traits are then adopted by breeders."

"That does sound much more risky than genetic engineering. Is that stuff actually in our food supply?" Amie exclaims, looking skeptically at the salad in front of us.

"Well, yes it is." I explain that such induced-mutation techniques are commonly used in traditional breeding and are thought to be quite similar to those that arise spontaneously in nature. Because the chemical dousing is done only once during the initial development of the mutant population and there are no chemicals left on the plants after several breeding generations, it is considered very safe. Breeders have taken advantage of both induced and spontaneous mutations to generate useful traits such as stress tolerance and improved grain characteristics. In the last seventy years, more than 2,250 mutant varieties have been released to plant breeders including rice, wheat, barley, grapefruit, and cotton; more than half of these were developed in the last twenty years (Ahloowalia et al. 2004).

"Sometimes mutant plants can be quite delicious." I say. Over 1,000 years ago, a spontaneous mutation gave rise to the rice needed for one of my favorite recipes, sticky rice with mango, an ancient treat from Thailand, which I share with the group now (box 7.1 and recipe 7.2).

BOX 7.1 The Rice *Waxy* Mutation

One technique to generate agronomically useful traits is called "mutagenesis." During this process, the seeds are put in a mutagenic and highly carcinogenic solution that induces random changes in the chemical letters of the DNA. The seeds are then planted and usually about 50% will die. The seeds from the

(continued)

BOX 7.1 *Continued*

surviving plants are collected, germinated, and surveyed for new traits by breeders. Such "induced-mutation" techniques have been commonly used in conventional breeding and can result in the same mutants derived from spontaneous mutations that occur in nature.

One of these spontaneous mutants gave rise to the precious sticky rice of Thailand. The sticky rice lacks the starch amylose, which constitutes up to 30% of the total starch in nonsticky rice endosperm. The lack of amylose is due to a mutation in a gene called *Waxy*, which encodes an enzyme required for amylose synthesis (Sano 1984). Sticky rice is an important culinary and cultural component throughout East Asia and is used in festival foods and desserts. In upland regions of Southeast Asia, it is a staple food in many homes. Ten percent of the rice traded each year is sticky rice.

The precise origin of "sticky" rice remains obscure because sticky rice is not found in the archeological record. Laotian Buddhist legend places the origin of sticky rice at around 1100 years ago, although Chinese folklore indicates that it was in existence around the time of the death of the poet Qu Yuan more than 2000 years ago (Xu 1992). A recent study by two North Carolina State University geneticists using modern genetic techniques now suggests that "sticky" rice most likely originated only once in Southeast Asia (Olsen and Purugganan 2002).

The researchers knew that different rice varieties all carry very similar genes, but in a significant proportion of cases (this varies from one species to the next) a given gene in one variety will be slightly different from its counterpart in a very closely related variety. That is the nature and basics of genetic diversity—genes differ in each rice variety. The researchers hypothesized that if a single breeder 2000 years ago developed a really good sticky rice variety, and then generously shared it with other breeders, the new varieties would all contain exactly the same *Waxy* gene. In other words, the *Waxy* gene would have a single origin. To test this idea they looked at the sequence of the *Waxy* gene of 105 rice strains and found that all those that were sticky carried nearly the same sequence in the *Waxy* gene, including a mutation that knocked out production of amylose. This result suggests that the early breeders of sticky rice liked the adhesive quality conferred by that single *Waxy* mutation and preserved that particular trait through breeding by incorporating it into new varieties with other desirable traits.

The development of "sticky" rice is a good example of how plant breeders choose modified plants in response to local cultural preferences.

RECISE 7.2

Sticky (Mutant) Rice with Mango or GE Papaya

The main ingredient in Thai sticky rice is rice carrying a mutation in the Waxy gene.

INGREDIENTS

1 lb Mutant rice (called "sticky" rice or "mochi" rice or "glutinous" rice)
1 Tbsp Salt
3/4 c Sugar
2 1/4 c Coconut milk
3 Peeled mangoes or GE papaya

1. Cover mutant rice with cold water and rinse. Repeat until the water runs clear, about three times, and drain.
2. Place rinsed rice in a bowl and fill with cool water so the water is approximately 2–3 inches above the rice. Let the rice stand in water for 6–8 hours.
3. Drain the rice, place it in cheesecloth, wrap it up, and put the cheesecloth inside a bamboo or metal vegetable steamer. Put 6–8 cups of water in steamer and bring to a boil. Cover and steam rice for 45 minutes (or until tender).
4. Meanwhile, dissolve salt and sugar in coconut milk, and heat—stirring to prevent lumps. When coconut milk mixture boils, stir on low heat until it is reduced to 1/3 of original volume.
5. Remove from heat and set 3/4 cup aside.
   Immediately after the rice is finished cooking place in a container with tightly fitting lid, and pour in remaining coconut milk mixture. Stir vigorously, cover, and let stand for 15 minutes.
6. Cut peeled mangoes or GE papaya into slices. Place on a serving plate. Spoon the cooked sticky rice beside the mango or GE papaya. Drizzle on the reserved coconut milk mixture. Serve and enjoy.

Matt jumps in, "But that case is different, a spontaneous mutation is natural. Surely that is a safer kind of rice than a rice mutant induced by carcinogenic compounds. In any case, we can avoid that risk by purchasing organic food."

I disagree. "This is a common misperception by many consumers who believe induced genetic changes are 'unnatural.' Under organic regulations, crops developed using chemical mutagenesis are acceptable and are not regulated for food or environmental safety."

Matt is surprised, "Even though the NAS reported that the mutation breeding process is *more* likely than genetic engineering to lead to unexpected consequences, mutant food can still be certified organic?"

"Yes, that is right. Risk is relative and this is a good example. Like all methods of breeding and genetic engineering, the risk of introducing a negative consequence by either method is extremely low. So low, in fact, that no one worried about unintended consequences until genetic engineering came along."

I pass around the rice-celery salad with purple tomatoes and ask Cindy what kind of rice she used (recipe 7.3).

She answers, "That is short grain brown rice from one of our favorite local companies, Lundberg Family Farms. I bought it at the Davis Food Co-op."

Pleased to have such a perfect example in front of me, I say, "This certified organic rice is derived from a radiation-treated variety called Calrose 76" (Ahloowalia et al. 2004; J. Jiang, Lundberg Family Farms, personal communication, 2006). The group looks at each other, seemingly surprised, but they all dig in nonetheless. Soon the crunching of celery fills the small room.

RECIPE 7.3

Rice-Celery Salad with Purple Tomatoes

INGREDIENTS

2 c Short-grain brown rice (derived from irradiation-treated Calrose 76) soaked
    in cold water for 30–60 min, washed, and drained
7–10 c Water
1 tsp Salt
1 Tbsp Olive oil

FOR THE DRESSING

2 Tbsp Chopped spring onions/scallions
3 Tbsp Chopped coriander/cilantro leaves
3 Tbsp Chopped mint
2–3 Tbsp Lemon juice
1 Tbsp Olive oil
Salt and pepper to taste
1 Cucumber, quartered lengthwise; then sliced thinly.
2–3 Stalks of celery, sliced thinly

FOR THE GARNISH

4 Pruden purple tomatoes, chopped

1. Heat the water in large saucepan and when it boils add salt and oil.
2. Add rice a little at a time (so water does not stop boiling).
3. Increase the heat a bit and cook rice for 10–12 minutes (stirring once or twice).
4. Strain rice into a colander, and as soon as water has drained, transfer rice to shallow tray, spreading and teasing it with a fork to remove any lumps.
5. Mix the dressing, including celery and cucumber, in a bowl.
6. DO NOT dress the rice while it's hot (heat will wilt the herbs).
7. Mix cooled rice and dressing in a large bowl and then transfer to a serving platter.
8. Garnish with tomato slices on the top or make a pile of chopped tomatoes in the center.
9. Refrigerate until needed. Bring to room temperature before serving.

Adapted from Sri Owen's "Rice Salad, Tabbouleh Style" in *The Rice Book*

In between bites, we talk about other compounds in food. Because plants are rich in sugars, proteins, vitamins and minerals, they make obvious and tempting treats for various predators. Plants cannot run away, so instead they have evolved a set of defenses to protect themselves. Celery is seemingly benign, yet it produces toxic compounds called psoralens to discourage predators and avoid being a snack too early in its life cycle. Sometimes humans are the accidental victims of psoralen poisoning.

Breeders have selected celery with relatively high amounts of psoralens because farmers prefer to grow insect resistant plants and consumers prefer to buy undamaged produce. Unfortunately, workers who harvest such celery sometimes develop a severe skin rash (NAS 2004), an unintended consequence of this conventional breeding. It is possible that if the gene encoding the toxic protein had been cloned and studied before being introduced into the new varieties, farm workers would have learned of the harmful effects before exposure.

Raoul thinks of another example, "I have discovered that green potatoes make pretty good rodent poison. One day I went into the certified organic hoophouse to find three dead mice near some freshly eaten green potatoes." Potatoes produce glycoalkaloid solanine, a toxic compound, although most varieties have amounts so small that they are considered nonhazardous to animals. Some potato varieties, however, have higher levels than others and certain conditions such as light can cause hazardous levels of the toxin to be produced.

So far, compounds that are toxic to animals have only cropped up in foods developed through conventional breeding approaches. There have not been any adverse health or environmental effects resulting from commercialized GE crops. This may be because foods produced by GE undergo additional scrutiny, or it may be that there simply are not yet many GE crops on the market. Whatever the reason, this important fact is sometimes lost in the debates on GE food.

"So what about traits from unrelated species? What if we put fish genes into rice, can the new trait itself be a problem?" Amie persists.

"Humans share enormous numbers of genes with plants and animals, so this particular example may or may not be problematic. The question, however, gets to the second important aspect of GE food. With GE, you can put genes from any species into a plant, which means that each new GE trait needs to be evaluated on a case-by-case basis." I reply. I explain that although there has never been a GE crop on the market that has harmed humans, there are cases of unexpected consequences in laboratory experiments. For instance, when a gene encoding a common bean protein was expressed in peas, a modified form was present that induced an immune response in mice. This GE pea was never commercialized (Prescott et al. 2005). Another example is Bt corn, which is engineered to be resistant to common corn pests (see box 5.3). Surprisingly, in addition to the expected resistance to earworm and rootworm, the transgenic Bt corn also contains significantly lower amounts of toxins produced by fungi that proliferate when stem are damaged by insects.

"Has any GE food on the market caused allergic reactions in humans?" asks Cindy.

"No," I reply. I go on to tell them about an experiment in which a known allergen (a protein from the Brazil nut) was engineered into soybean. The new GE variety induced production of reactive antibodies in human sera of in individuals previously known to be allergic to Brazil nuts (Nordlee et al. 1996). This variety of GE soybeans was never field-tested nor commercialized for chicken feed as originally intended, partly because comprehensive safety evaluation of this GE crop revealed this adverse effect prior to commercialization.

As a midwife, Amie is well-versed in risks associated with pregnancy and childbirth. She reminds me that when I was pregnant with our third child, Audrey, she suggested that we deliver Audrey at our home in our outdoor hot tub. "Too risky," many of our friends and family counseled. Our French friends, Serge and Evelyne, were especially surprised, because in France, it is more common to drink wine during pregnancy than to give birth at home in a hot tub.

The prevailing assumption is that hospital-based deliveries are safer for both mother and child; indeed most women do prefer the comfort of knowing that the most modern technology is close by, especially because there are real and documented hazards clearly associated with childbirth. Yet, studies have shown that healthy women who wish to deliver at home have no increased risk either to themselves or to their babies (Ackermann-Liebrich et al. 1996). In fact, at least one study showed that for women who have previously given birth and have planned ahead, delivery at home with a trained midwife is actually less risky than a hospital birth (Wiegers et al. 1996). These scientific studies were reassuring, and because we lived five minutes from the hospital and knew that Amie's skills were superb, we decided to deliver Audrey at home.

Amie smiles, she seems to know what I have been thinking, "It turned out okay, didn't it?" Indeed, she is right. Audrey was born at high noon on a spring day in the water under the perfume of our trellised purple wisteria. I remember clearly the sun hats on all of us, the laughter, and the sweet taste of the freshly squeezed lemonade my mother had lovingly prepared.

"But really, it's all about the children for me," Amie continues in a serious tone. "I don't mean only our own children, but all those who will grow up on this planet. We want to leave a cleaner environment that will support abundant and nutritious crops."

"Uh-huh," I agree. I am not able to say much more at the moment because my mouth is full of Christollen. Yet, Amie's comment gets me thinking about the potential impact of genetic engineering on young children in the developing world and our responsibility to them.

It is well established that vitamin A deficiency (VAD) is a public health problem in more than 100 countries, especially in Africa and Southeast Asia, hitting hardest young children and pregnant women. Worldwide, over 124 million children are estimated to be vitamin A deficient. Many of these children go blind or become ill from diarrhea, and nearly eight million preschool age children die each year as the result of this deficiency. The World Health Organization estimates that improved vitamin A nutritional status could prevent the deaths of 1.3–2.5 million late-infancy and preschool age children each year (Humphrey et al. 1992). The heartache of losing a child to a preventable disease is not one commonly encountered in the developed world.

To combat VAD, the World Health Organization has proposed an arsenal of nutritional "well-being weapons" including a combination of breastfeeding and vitamin A supplementation, coupled with long-term solutions, such as the promotion of vitamin A-rich diets and food fortification. In response to this challenge, a group of Rockefeller Foundation-supported scientists decided to try to fortify rice plants with higher levels of carotenoids, which are precursors to vitamin A. They introduced a gene from daffodils (which make a lot of carotenoids, the pigment that gives the flower its yellow color) and two genes from a bacterium into rice using genetic engineering (Ye et al. 2000). The resulting GE rice grains were golden and carotenoid-rich.

Amie refills our cups and I take a sip before continuing. "In a sense, the resulting nutritionally enhanced rice is similar to drinking vitamin D enriched milk—except put there by a different process."

"Yes," I reply. "It is also similar to adding iodine to salt; a process credited with drastically reducing iodine-deficiency disorders in infants."

"Will eating carotenoid-fortified rice really help the children?" Matt asks.

I tell them about my recent conversation with my colleague, Mike Grusack who works at the Children's Nutrition Research Center in Houston, Texas. Mike told me that preliminary results from human feeding studies suggest that the carotenoids in

"Golden Rice" can be properly metabolized into the vitamin A that is needed by children (M. Grusack, personal communication, 2006). Other studies also support the idea that widespread consumption of Golden Rice would reduce vitamin A deficiency, saving thousands of lives (Stein et al. 2006). The positive effects of Golden Rice are predicted to be most pronounced in the lowest income groups at a fraction of the cost of the current supplementation programs (Stein et al. 2006). If confirmed, this relatively low-tech, sustainable, publicly funded, people-centered effort can complement other approaches such as the development of home gardens with vitamin A-rich crops such as beans and pumpkins.

Unfortunately, because vitamin A rice is the product of genetic engineering, some people view it with suspicion and worry about long-term consequences (box 7.2). Similarly in some nations, iodization was thought to be a governmental plot to poison the salt. In a 2006 *New York Times* article, journalist Donald McNeil describes how iodized salt was blamed for AIDs, diabetes, seizures, impotence, and peevishness. "Iodized salt...will make pickled vegetables explode, ruin caviar or soften hard cheese." In Kazakhstan, breaking down that kind of resistance took both money and political leadership. But it eventually succeeded. Today 94% of households in Kazakhstan use iodized salt and the United Nations is expected to certify the country officially free of iodine-deficiency disorders.

BOX 7.2 **Indicators of Long-Term Health Risks**

Even though the vast majority of scientists view the process of GE as safe, some consumers still worry that the consumption of GE food could cause long-term effects that we will not determine until it is too late.

This is just the kind of concern that is addressed by Allan Mazur, a sociologist and professor of public affairs at the Maxwell School of Syracuse University, in his book *True Warnings and False Alarms*. To identify hallmarks that could help predict the truth or falsity of an alleged hazard, Mazur assesses 31 health warnings raised between 1948 and 1971 about diverse technologies including pesticides and fluoridation of community drinking water. With 30–50 years of hindsight, he identifies three characteristics, apparent from the outset of a controversy that most effectively distinguished true warning from false alarms.

1. Warnings turned out to be more than twice as likely to be true if the first conspicuous source of the public warning was based on a report of scientific research produced at a recognized scientific institution. If the alarm was raised by a government agent or citizen advocacy group it was more likely to be false.

2. True warnings were less likely than false alarms to have sponsors with biases against the producer of the alleged hazard.

(continued)

BOX 7.2 *Continued*

---

3. Warnings appearing in the news partly because of their connection to earlier news stories were more often false than warnings reaching the news without a boost from collateral sources (Mazur 2004).

It has been nearly thirty years since the first warnings were raised about the process of genetic engineering as being inherently dangerous. Here, I apply Mazur's criteria to assess the safety of GE food.

1. The first conspicuous source of a public warning about releasing GE organisms into the environment was raised by a citizen advocacy group. In my second year as a graduate student at UC Berkeley, there was a public outcry over the first Environmental Protection Agency approved release of genetically engineered organisms into the environment. The bacteria, called Frostban, was created to protect plants from frost by Steve Lindow, a low-key professor who worked across the hallway from my laboratory. The early-morning application of Frostban on a strawberry patch in 1987 was witnessed and reported by 75 media outlets from throughout the world (Kahn & Co. 2006). This step inaugurated both the field use of genetic engineering for agricultural and environmental purposes, as well as the beginning of protests against the emerging biotechnology industry. Although Frostban was widely viewed as harmless to the environment and potential consumers by the scientific community, critics warned of possible environmental disasters from this "unnatural process" (Rifkin 1983). Today, there is no longer any significant concern about this product. In fact, a related strain is being used at ski resorts under the trade name Snomax to increase the effectiveness of their snow-making machinery (York Snow, Inc. 2001).

2. Many of the sponsors of anti-GE protests dislike Monsanto, one of the first large scale commercial producers of GE seed. These critics are concerned that the GE process will enhance corporate control of our food supply. Such increased control, however, if it occurs, would be socioeconomic and have little to do with the risk of eating the food. Such nonscientific concerns can only be addressed through policy.

3. The warnings about GE food coincided with outbreaks of food safety problems in other parts of the world. For example, mad cow disease (which has nothing to do with GE) was first reported in 1986 in the United Kingdom (UK), shaking confidence in the reliability of regulatory agencies.

All three of these characteristics are typical of false alarms. Therefore, the belief of the vast majority of scientists and scientific organizations that no long-term or short-term risks are likely to be associated with the process of genetic engineering is also supported by Mazur's criteria.

Raoul changes the subject to ask about the safety of the antibiotic resistance genes that are sometimes present in transgenic plants. "What if the antibiotic resistance genes are somehow acquired by the bacteria that live in our intestines?"

"According to a committee of the National Academy of Sciences, this is unlikely" I answer. First the gene would have to escape the human digestive juices, then it would have to survive intact in the human gut and finally it would have to move into the intestinal bacteria. Indeed, one study showed that transgenes in GE soy are completely degraded by the time they get to the large intestine (Netherwood et al. 2004). Furthermore, many antibiotic resistance genes are already common in bacteria and have been in our food all along. There are also several technological advances that make Raoul's concern even more remote. For example, new markers, such as sugar enablement markers (see figure 4.3) are now available, so antibiotic resistance genes are being used less often. Also, many new transgenic crops, such as XA21 rice that is resistant to bacterial disease, do not contain marker genes at all (Dr. Shirong Jia, Biotechnology Research Institute, Chinese Academy of Agricultural Sciences, personal communication). A far greater risk is the overuse of antibiotics for medical purposes, which selects for resistant bacteria (Gold and Moellering 1996) and the use of animal feed supplemented with antibiotics (Kidd et al. 2002).

Hens are a prime example of the risks of using antibiotics in feed. In most large commercial poultry operations, low levels of antibiotics are mixed with their grain, both to spur faster growth and to counteract the spread of disease that stress and living in close quarters (more than twelve birds per square meter) can promote (Food and Drug Administration Center for Veterinary Medicine 2000). It is now known that after repeated exposure to low doses of antibiotics, bacteria can become resistant. If antibiotic-resistant bacteria proliferate in the intestines of the hens and if people then eat undercooked chickens, people can become ill. Scientists now hypothesize that antibiotic use is linked to the evolution of multiple drug-resistant bacteria and the loss in efficacy of drugs important to human medicine, though this is difficult to demonstrate by rigorous scientific methods (Anderson et al. 2003; Committee on Drug Use in Food Animals, Panel on Animal Health, Food Safety, and Public Health, Board on Agriculture, National Research Council 1999; Wiley et al. 2004).

Amie says, "Even if GE crops are considered safe by most scientists, why not simply label the produce from these crops and let people decide for themselves? I like to know what I am eating and make my own choices."

I answer, "I am also a label reader. If there is an excess of added sugar or too many ingredients with names that I don't recognize then I don't buy the product. Not all information, however, is useful."

A few months ago our local food coop began posting red "consumer alert" signs that say, "Conventional foods that contain corn, soy, or canola may be genetically engineered." I find these signs more annoying than helpful. It is a little bit like the warnings posted on science textbooks in some states that say, "This textbook discusses evolution, a controversial theory which some scientists present as scientific explanation for the origin of living things, such as plants and humans. No one was present when life first appeared on Earth. Therefore, any statement about life's origins should be considered as theory, not fact" (Oklahoma Legislature 2003; Cline 2006). Neither statement says anything informative about the state of our food nor the creation of our universe. With no specific hazards associated with GE foods or evolution, how can a consumer use these statements to make a more informed choice about the risk to their health or to their faith in God?

The National Research Council Committee states that attempts to assess food safety based solely on the process are scientifically unjustified. Rather than adding a general label about the process with which a plant variety was developed, it would make more sense to label food so that consumers are informed about what is actually in or on the food. But this, too, is not necessarily helpful. For some people it may be informative to read a label that says, "may contain traces of carbamate pesticides, which at high concentrations are known to cause death of animals" or "may contain trace amounts of purified *Bacillus thuringiensis* protein, which kill Leptidoptera (a class of insects)." But is it helpful to most consumers who are not familiar with the science?

Here is another example. If we carry forward with labeling the product, then organic produce treated with rotenone, a "natural" pesticide favored by some organic farmers, would need to be labeled with the following, "may contain trace amounts of rotenone—chronic exposure can cause damage to liver and kidney" (Occupational Safety and Health Administration 1998). Organic super sweet corn would require this label: "Carries a genetic mutation induced by radiation mutagenesis, resulting in the presence of a mutant protein." Organically grown papaya would need to be marked: "may contain vast amounts of papaya ringspot viral RNA and protein" (see box 4.2). These labels are so ominous that it is not likely that many people would feel comfortable eating these organic fruits and vegetables. Still, there is no evidence that any of these food products are hazardous. After all, we have been eating sweet corn and organic papaya safely for years.

"It seems to me that if the labeling statement does not help with safety interventions or inform consumer choice, it does not serve the purpose" I say, finally answering Amie's question. "It only confuses and unnecessarily alarmspeople."

As it is almost time to go, we pause for a last snack of cheddar. I mention that the cheeses we are eating, like most that have been sold since 1990, were made with genetically engineered ingredients. Before that, the curdling agent was isolated from the stomach of young calves (see box 5.1).

Matt looks interested and remarks, "The good thing about it is that you can eat GE cheese and still keep kosher." Matt follows this Jewish tradition, which prohibits eating meat and dairy together.

Amie, a vegetarian, wrinkles her nose and says jokingly, "How awful—enzymes from slaughtered animals. I prefer the GE cheese, but, Pam, I think I will let you take the first bite."

On that note, we gather our plates, bring them to the kitchen, step out the door, and say good-bye.

The next day, I am in my office sitting restlessly in front of my computer, fielding too many questions over e-mail. I feel overworked and short on time. A quick glance at the clock tells me that I am late again for swim practice so I grab my bag, skip downstairs, step outside, jump on my bike, and pedal to the pool. Usually swimming in the middle of the day alters my state of mind dramatically.

Five minutes later, I pull up to the chain link fence surrounding the pool and park my bike. My swim buddies are talking and laughing; they look comfortable in their bathing suits even though it is January. I hurry into the locker room, change into my swimsuit, pull on my plastic cap and press the goggles snugly over my eyes. I head out to the pool and jump in. It is cold so I swim quickly to warm up. Soon I am listening to the chugging sound of each stroke and starting to forget the business of the laboratory. When I come up to breathe there is a faint smell of chlorine and coconut sun-screen. In Davis, every day is UV-rich and swimmers' skin needs protection. At each end of the pool, I flip, feel for the wall and push off with head squeezed between my arms, toes pointed, and body like an arrow. I enjoy the feeling of slicing through the water.

As I swim, I think about the definition of risk used by David Ropeik, director of risk communication at the Harvard Center of Risk Analysis: "the probability that exposure to a hazard will lead to a negative consequence" (Ropeik and Gray 2002). Overexposure to both sun and water can cause death from skin cancer or drowning. Yet it is a risk we take because the benefits are clear. But what if it were explained in a different way? Take water for example, which is composed of two atoms of hydrogen and one of oxygen. Saying "water" does not frighten people but what if they heard someone say "Dihydrogen Monoxide (DHMO) (i.e., water), a colorless and odorless chemical compound, is perhaps the single most prevalent of all chemicals that can be dangerous to human life. Despite this *truth*, most people are not unduly concerned about the dangers of Dihydrogen Monoxide" (Way 2006, emphasis in original). With such a warning and without additional information, I would not go near the stuff.

Both sun and water are essential to human life, as well as to the plants we depend on for nutrition. One could argue that breeding and genetic engineering are not essential, but I can't think of any place on Earth that does not rely on crops that have been bred for improved characteristics. Some will argue that we should simply stick with standard breeding because so far, the negative consequences have been minimal. Somehow this argument discounts the thousands of deaths or poisoning each year from pesticide exposure and the crop losses that have led to starvation. As far as we know, there have been no negative consequences associated with genetic engineering. The probability of exposure to a possible hazard is quite low for GE crops, because of the additional regulatory requirements. These are some of the reasons that Roepeik ranks GE foods as low risk (Ropeik and Gray 2002).

One hour later I pull myself out of the water feeling rejuvenated, young again, and as if I have all the time in the world.

As I bicycle back to my office, I think about the fact that some people eat food known to be a health hazard, because the benefits seem to outweigh the risks. This is certainly true for me. A few days ago, when I was preparing the Christollen egg-bread, I could not resist tasting the rich, buttery dough. Similarly, I cannot resist dough from my favorite cookies (recipe 7.4).

RECIPE 7.4

Pam & Trish's Oatmeal Chocolate Chip Cookies

INGREDIENTS

1/2 c Safflower oil
1/2 c Unsweetened butter
1 c Brown sugar
2 Eggs
1 tsp Vanilla
2 c Freshly ground oatmeal (use food processor or blender)
1 c Whole wheat pastry flour (as fresh as possible for both flours; we use Jennifer's Windborne farm flours)
1 c Barley flour
1 tsp Soda
1 tsp Baking powder
1/2 tsp Salt
1/2 c Wheat germ
3/4 c Raisins
1 1/2 c Chopped walnuts (as fresh as possible...we use Terra Firma organically grown)
1 c Grain-sweetened chocolate chips

1. Preheat oven to 350°F.
2. Beat together safflower oil, butter, and brown sugar until fluffy. Beat in eggs and vanilla.

3. In a separate bowl, mix together ground oatmeal, pastry flour, barley flour, soda, baking powder, salt, and wheat germ.

4. Stir oatmeal mixture into egg mixture. Mix in raisins, walnuts, and chocolate chips.

5. Drop by large spoonfuls onto ungreased cookie sheets and bake for about 10 minutes. Warning—tasting cookie dough may be hazardous to your health!

I remember my mother making chocolate chip cookies after school, and the arguments with my brothers over whose turn it was to lick the beaters or bowl. Nowadays when I make cookies with the children, we stick our fingers in the soft, gooey batter and let the sweet stuff melt in our mouths (and then usually return the finger to the bowl again, another ill-advised health practice). So far none of us have become sick.

Raoul reminds me often that eating uncooked egg dough is hazardous, and he is right. A bacterium, *Salmonella enteritidis,* can be inside perfectly appearing eggs, and if the eggs are eaten raw or undercooked (even if they are gathered from your own hens as we usually do), the bacterium can cause severe illness. The Centers for Disease Control estimates that 1 in 50 consumers are exposed to a contaminated egg each year (CDC 2005). Still, because it is a family tradition and a delectable pleasure to eat raw dough, I take the risk.

I recall the only time I was able to resist the lure of the raw cookie dough—when I was pregnant with our first child. Nor did I eat raw milk cheeses, certain fish, ready-to-eat meats, or alfalfa sprouts—all foods known to pose risks during pregnancy. I chose only the safest foods to nourish our baby. We even checked his chromosomes through amniocentesis, which thankfully revealed a perfect child—a healthy boy. None of the many precautions we took protected Ivan from the twisted cord, the cutting off of oxygen, the premature death that left us bereft, empty and unable to communicate with ordinary human beings.

Fortunately, Amie is extraordinary, and she was on call the day Ivan was stillborn. I asked her how one could survive such pain. She looked me straight in the eyes and did not turn away from the terror and desolation that must have been apparent. She asked everyone to leave the room and crawled onto my hospital bed and answered my question. She described unexpected losses in her own life and said that meditation was her comfort. I was not convinced that day that there would be a way through this time for me. But I did learn that there was an exceptionally kind and compassionate person in the world, and for the moment that was enough.

We struggled to understand this random death. We had not been aware of the risk of cord accidents; a risk that seemingly runs as high as 1% for every healthy pregnancy. It is not understood why—only that the nutrition and activity of the mother are not factors. It is thought to simply be a tragic fluke—the baby twisting one way and the cord another. It is a risk we take in our desire to have a child.

For most humans, all the essentials of life—food, family, and work have associated risks. We choose our risks consciously or unconsciously. The greatest hazards usually arise from known high-risk behaviors such as smoking or overeating, but sometimes they arise from some randomness we could not have foreseen. In the case of pregnancy and childbirth, the development of modern medical technologies have greatly reduced the risk of stillborns but not eliminated them. In the case of food, there is always some risk when we eat, whether the food was generated through breeding or genetic engineering.

In the end, we can only gather the most accurate information from reliable sources and make the best choices possible. I know that the GE crops currently on the market are no more risky to eat than the rest of the food in our refrigerator. And the same technology has a significant potential for saving children's lives, whether through reduced exposure to pesticides or increased nutrients in their diet.

As for Amie, I think she will continue to let me take the first bite of any new genetically engineered product on the market. For our friendship I will do it. I wouldn't risk that for anything.

The Environment

Eight

CONSERVING WILDLANDS

Agri cultura . . . est scientia quae sint in quoque agro serenda ac facienda, quo terra maximos perpetuo reddat fructus. (Agriculture is a science, which teaches us what crops are to be planted in each kind of soil, and what operations are to be carried on in order that the land may produce the highest yields in perpetuity.)

MARCUS TERENTIUS VARRO, a Roman landowner
of the first century B.C., as quoted in Sir Gordon
Conway, *The Doubly Green Revolution*, 1997

It is 10 A.M. on a wintry November morning. The children, Raoul, and I have hiked nearly an hour from Emerald Bay in the Sierra Nevada Mountains to reach the top of this ridge. To get here, we passed through a narrow canyon that was cold and shaded by steep granite walls, the early morning sun still hidden behind 9,000-foot peaks. Only now, as we look down upon Eagle Lake, has the sun begun to warm us. The children see the water below and eagerly run ahead. We follow, make ourselves comfortable on some logs, and eat our lunch. A great diversity of tree species tower above us, including white firs, sugar pines, incense cedars, and an occasional red fir. Many of these trees escaped the logging of the 1850s and have now lived longer than the reach of our parents' memories. Mountain chickadees and Steller's jays dart through the trees. It is very quiet; we feel as if we are the only ones awake in the world, and the only ones to know this place. The beauty and wildness here seeps into our bones as the tensions of our scheduled lives dissolve.

As the world's population grows (it is expected to increase from the current 6.7 billion to 9.2 billion in 2050), fewer and fewer of these wild places remain (Population Division 2007). Today vast areas of Earth resemble the agricultural Central Valley, where a few domesticated species dominate. As the demand for food increases, will it be possible to minimize the impact of food production on what remains of wild nature?

We gaze at an island in the lake, the lone tree upon it glistening with frozen droplets of water. The shore is covered with black ice that can be distinguished from the water only by the way it amplifies the bright morning sunlight that is now pouring over the ridge.

"Oh, what is it?" six-year-old Cliff suddenly exclaims, his blue eyes wide and focused in front of him. We all turn and see on the lake a flash of light like a star or a firecracker exploding, then another and another until there are hundreds of patches of twinkling stars improbably skimming the surface of the water, moving from one place to the next in a rhythm that is their own, impossible to predict.

We edge closer. What a strange feeling this scene evokes; this almost supernatural dancing of light on water as if fairies of the wilderness were visiting us from a faraway world. The 100-mile drive from Davis, the smells of the musty car, the small anxieties and noise of our everyday lives, the monotony of monoculture, all simply evaporate in a flash of awe and speculation.

"It is my magic," four-year-old Audrey states matter-of-factly. "I let it out to play." We are startled, because before today we had not seen evidence of such special powers.

"Nuh-uh," Cliff argues shaking his head. "There is no such thing as magic. It is not for real, right, Mama?" He looks at me hopefully, wanting to refute this particular hypothesis that seems to favor his sister.

"Well," I answer, "it seems magical to me although I can't say for sure, because I have never seen anything like it."

Satisfied and in high spirits, they go off to stomp on patches of ice, their blond and brown heads bouncing up and down. They jump up on top of a fallen log that extends into the deep part of the lake, and begin to explore in that direction, Raoul trailing at a safe distance.

⸺⸰⸺

Left alone to lounge in the sun, I walk to the edge of the lake and with my hands cupped, draw the water to my mouth; its purity and sweetness are unmatched. This taste is as much a part of my childhood as are the scent of the pine trees and the rough texture of the granite. As I listen to the delighted cries of the children encountering the wildness and beauty here, I think of the importance of preserving these vanishing places where the pristine dominates and humans are appreciative visitors rather than consumers. Both the young and the old benefit from wild nature in many ways: nature rejuvenates the spirit, regulates our climate, cycles nutrients, and provides scientists with sources of genes needed for development of new crop varieties and drugs. One study estimates that the overall economic benefit-to-cost ratio of an effective global program for the conservation of remaining wild nature is at least 100 to 1 (Balmford et al. 2002).

These benefits should act as powerful incentives to conserve what remains of natural ecosystems; unfortunately, however, loss and degradation of natural habitats continue largely unabated. The most devastating impact on biodiversity result from global agricultural activities cause (Schaal 2006). In fact, farming is already the greatest extinction threat to birds, and its adverse impacts look set to increase, especially in developing countries (Green et al. 2005). I am relieved that this land is not good for farming because if it were, it would almost surely be destroyed.

Raoul and the kids are ready to hike back. I reluctantly pack up, and we climb the short distance up to the ridge and then make our way down the canyon, passing a few other hikers on our way out. On our drive home, as we descend from Echo Summit down into the valley, I feel keenly, as I always do, the separation from the mountains.

In less than an hour and a half, we pass over the last foothill of the Sierra Nevada Mountains and begin to drop down into the Central Valley. Straight ahead, fifty miles to the west is the Coast Range. Millions of years ago, erosion from these two parallel mountain ranges deposited sediments onto the floor of the valley that was once an inland sea, creating soil that is deep brown, loamy, and rich in nutrients. Today this fertile soil supports the most productive agricultural land in the world, supplying 25% of all the food produced in the United States.

The valley was not always planted to crops. One hundred and fifty years ago grizzly bear, Tule elk, pronghorn antelope, coyote, and deer prowled through riparian forests that bordered the great rivers of the Sacramento and San Joaquin and their tributaries. Indeed, the first description by Spaniards of the Central Valley in July 1769 suggests an earthly paradise: "the place where we halted was exceedingly beautiful and pleasant, a valley remarkable for its size adorned with groves of trees and covered with the finest pasture" (Brandes 1970).

In the mid-1800s, the agricultural promise of this valley and the discovery of gold in the nearby foothills drew people from all over the world. The valley was grazed, burned, and plowed. The subsequent, explosive expansion of grain farming transformed the Central Valley and its residents. As a result, farms and cities flourished whereas most of the wild animals, thick forests, and valley bunchgrasses vanished, and all but one of the rivers that drain from the Sierras were dammed. At dusk, as if they were calling for the wildness to return, coyotes' howls can occasionally be heard.

I can relate to their call. Not long ago, on my spring bicycle rides to campus, I was able to appreciate the vibrant view of remnants of the once-extensive valley gardens. Situated between the bike path and a farm near our house were vernal pools surrounded by a bright tapestry of blossoms: goldfields, purple owl clovers, and blue lupines. Such pools are found on impermeable layers of ancient soils throughout the valley; they do not drain well and are marginal for farming. Each rainy season the pools would fill up and serve as a home for fairy shrimp, freshwater insects, and frogs. Birds visited to feed on the vernal pool's plants and animals. As the pools slowly dried out each spring, brightly colored concentric rings of flowers would appear.

One day the pools were gone. The hardpan had been broken up with enormous chisels pulled by huge tractors. The farmer then planted wheat, in order to eke out a little bit more yield on the same acreage. Such expansion of farming to pristine areas occurs all over the world, destroying vast quantities of wilderness and its associated wildlife each year (Green et al. 2005).

As we drive through the valley passing new developments, farmland and ranchland, I hand out snacks to my hungry family. "Tofu again?" they complain. I realize that, yes, it is again tofu, diced into cubes for this road trip.

Not much has changed in my diet, since I became a vegetarian thirty years ago. Concerned then, as I am now, about the need to feed a growing population, and influenced by Frances Moore Lappé's book, *Diet for a Small Planet*, I quit eating meat. The idea was that if everyone on Earth converted from a meat-eating diet to vegetarianism vast areas of land would be freed from cultivation (Lappé 1971). This is because much of the Earth's agricultural land is dedicated to grain production to feed the growing number of animals that consumers demand. Unfortunately, despite my good intentions and those of many like-minded people, the trend has been in the opposite direction worldwide; even developing countries are increasing their meat consumption (Delgado et al. 1999; Myers and Kent 2003). Global livestock grazing and feed production now use 30% of the land surface of the planet, destroying much biologically sensitive terrain in the process (Steinfeld et al. 2006).

Reducing population growth would be another way to address the problem of diminishing wildlands. After all, fewer people on Earth would mean that less land needs to be used to grow food. But who decides how to go about this? Some countries oppose birth control and others are encouraging their citizens to reproduce, even providing incentives to do so (Chivers 2006). Clearly, there are a host of social, economic, and political issues that mediate the relationship between agricultural production, human nutrition, and habitat protection.

Despite these complexities, there are still immediate actions we can take to reduce the negative ecological impacts of farming. The first is to maximize wildlife-friendly practices. Part of this approach aims at retaining patches of natural habitat on farms to provide shelter for wildlife (Green et al. 2005; Rosenzweig 2003a). Unlike the high-elevation mountain ranges where only the most dedicated gardeners can grow edible crops, the most productive agricultural land is usually found in warm, sunny areas with sufficient water and deep soil, and tends to support a great diversity of animals, amphibians, and other creatures. In Costa Rica, for example, half of the native forest species of birds, mammals, butterflies, and moths live near highly productive farmland. Therefore, retaining patches of natural habitat on these farms benefits the neighboring wildlife (Green et al. 2005; Rosenzweig 2003b).

There is one caveat to this approach to biodiversity conservation: if the farms do not produce well, farmers are forced to use more land, thus reducing the land available for wildlife habitat. How, then, can farmers cultivate wildlife-friendly farms that retain biodiversity and at the same time provide enough food for a growing and increasingly demanding human population? The Ecological Society of America suggests that maintaining high yields per acre will be critical (Snow et al. 2005). Such an approach would reduce pressure on natural habitats, because less area would need to be cultivated for a given amount of yield.

Over the last century the most efficient, economical and environmentally-friendly approach to increasing yield has been through plant breeding. Barbara Schaal, an ecologist at Washington University and a member of the National Academy of Sciences, argues that plant breeders have done more for conservation of biodiversity than any other group of humans (Schaal 2007). Many scientists agree. Rhys Green, a biologist at the University of Cambridge, and his colleagues point out that without the development of high-yielding crop varieties over recent decades, two to four times more land would have been needed to produce the same amount of food in the United States, China, and India. Looking ahead, they calculate that without additional yield increases, maintaining current per capita food consumption will necessitate a near doubling of the world's cropland area by 2050. By comparison, raising global average yields to those currently achieved in North America could result in very considerable *sparing* of land (Green et al. 2005; Waggoner 1995).

Clearly, genetic modification through plant breeding has been critical to increased land sparing in the past and will continue to be so in the future. But production practices are also important for the ecology of the land. For example, much of the high yield achieved in North America is dependent on synthetic inputs such as pesticides and fertilizers, which are costly and can degrade the environment, reducing biodiversity. It is estimated that the pesticides used in the United States kills seventy million birds each year as well as billions of insects, both beneficial and harmful. Such environmental losses cost the public about $1 billion each year (Pimentel and Raven 2000). Herbicides, used to kill weeds, also have negative impacts. Atrazine, the most commonly used herbicide in the United States and probably the world, causes male demasculinization and hermaphrodism of African clawed frogs. Overuse is speculated to be responsible for the drastic reduction in frogs worldwide over the last fifty years (Hayes et al. 2002). The Global Amphibian Assessment found that nearly one-third of the world's 6,000 or so species of frogs, toads, and salamanders face extinction—a figure far greater than that for any other group of animals (GAA 2004). These examples illustrate one of the global challenges for the next century: the need to develop high-yielding varieties that require minimal inputs, so that impacts on biodiversity can be minimized.

An alternative to the "high-input" approach is to expand the number of organic farms. Because organic farmers do not use synthetic pesticides, their farms support higher levels of biodiversity than conventional farms. Furthermore, there is accumulating evidence that organic farming can yield as much, for some crops, as conventional agriculture (Reganold et al. 2001; Mader et al. 2002). Unfortunately, because the biodiversity value of farmland generally declines with increasing yield on a given piece of land, even organic farms usually host far fewer species than do original pristine ecosystems (Green et al. 2005; Pain and Pienkowski 1997; Krebs et al. 1999; Donald et al. 2001). What this means is that even if we convert ALL of agriculture to organic farms (now only 2% in the United States), we still need to increase yield if we want to spare land and protect wildlife.

Fortunately organic farmers in the U.S. are usually looking for ways to create more value per hectare and one of the best ways is to increase yield (Guthman 2004). The most economically successful commercial organic farm operations tend to be quite intensified, producing very high yields. This was certainly true for Full Belly Farm, the organic farm that Raoul helped establish many years ago, as well as at the U.C. Davis Student Farm where he is farming now.

I nudge Raoul from his road-induced stupor with a question, "How many crops do you grow on your land each year at the student farm?"

"Three crops, including a cover crop," he replies. "I typically plant lettuce in the early spring, followed by tomatoes or eggplants, and then a cover crop."

This type of crop rotation and intensification is good for the wildlife on the farm and for sparing land *from becoming* farmland, which is the greatest benefit to wildlife. Both conventional and organic farmers rely on genetically diverse and improved plant varieties to increase their yields, and conventional genetic modification is the basis of all such improved varieties. GE has lead to even more new tools to for breeding pest resistant crops, thus reducing the application of pesticides on conventional farms and increasing yield (Fernandez-Cornejo and McBride 2000). For this reason, some ecologists see the application of GE as a way to spare even more land from destruction by enhancing yields (Qaim and Zilberman 2003; Snow et al. 2005).

In China, the results are dramatic: an 80% reduction in pesticide use on small farms planted with GE insect-resistant cotton (Huang et al. 2005; see box 5.3). Similarly, the USDA Economic Research Service reports that pesticide use on corn, soybeans, and cotton declined by about 2.5 million pounds in the United States since the introduction of GE crops in 1996 (Fernandez-Cornejo and Caswell 2006). These results support the notion that GE may radically reduce the negative impacts of farming practices on the environment and spare more land for wildlife (figure 8.1).

Interestingly, even GE plants that were designed to be used with herbicides have helped eliminate the application of more toxic herbicides, which, in turn, has led to greater biodiversity on farms (Strandberg and Pederson 2002). This is because the GE herbicide-resistant plants are used with newer broader spectrum herbicides such as glyphosate that target only plant metabolic processes and do not persist in groundwater (see box 5.2). Conversely, atrazine has been detected in the ground water and

FIGURE 8.1 Save the Earth T-Shirt.

rain of most mid-western states (Durkin 2003; Giesy et al. 2000). Rachel Long, a UC Cooperative Extension Adviser in Yolo County and a member of the Organic Farming Research Workgroup, tells me that currently most conventional alfalfa farmers in the Central Valley use diuron to control weeds. Like atrazine, diuron also persists in ground water and is highly toxic to aquatic invertebrates (Extension Technology Network 1996). "I am hoping that the new GE herbicide-resistant alfalfa variety just developed by Monsanto will help improve water quality in the valley," she told me recently.

Water conservation also requires high agricultural productivity. Water systems are under severe strain in many parts of the world. Many rivers no longer flow all the way to the sea; 50% of the world's wetlands have disappeared and many major groundwater aquifers are being mined unsustainably, with water tables in parts of Mexico, India, China, and North Africa declining by as much as one meter per year (Somerville and Briscoe 2001). Approximately 40% of the world's food is produced from irrigated land, and 10% is grown with water mined from aquifers. Thus, increased food production must largely take place on the same land area while using less water. More-effective management of water requires a series of institutional and managerial changes in addition to a new generation of technical innovations that includes advances in genetic engineering of plants for drought tolerance and pest resistance (Somerville and Briscoe 2001). Reducing losses to pests and pathogens is equivalent to creating more land and more water because such losses account for an estimated 40% of plant productivity in Africa and Asia, and about 20% in the developed world. According to Chris Somerville, Director of the Carnegie Institute of Washington at Stanford and a member of the National Academy of Sciences, "The benefits of genetically engineering new crop varieties with increased pest resistance, drought tolerance and higher water use efficiency would be substantial in terms of income and food for the poor, reduced demand for water, and limiting the expansion of land area under cultivation, all of which would also generate environmental benefits."

We are almost home and the children are restless in the car.

"Mama, when are we going to be there?" Cliff whines charmingly.

To distract him I ask, "Do you want to listen to James and the Giant Peach?"

"Yeah!!!"

I push in the cassette. As we listen to the weird and imaginative prose of Roald Dahl, I start thinking about the peach that grew and grew following its contact with the magic green wiggly things, becoming sweeter and juicier as it plumped up. For weeks, this single peach nourished a group of diverse creatures, in the close quarters of a hollowed-out pit, consisting of worms, flying insects, a spider, and a small boy named James. I wonder what Roald Dahl, who wrote this book in 1961, would think about the progress in plant biology, genetics, and the resulting seemingly infinite possibilities for crop alterations. I imagine he would find it a great source of material for another delicious novel.

We arrive in Davis and stop at the student farm. Audrey and Cliff tumble barefooted from the car, happy to run freely. As Raoul waters the seedlings in the greenhouse (lettuce, kale, collard greens, chard, parsley), the children and I wander through a wooden gate and out to the orderly rows of vegetables, where the remnants of unharvested pumpkins lay about, their brightness acting as magnets for the crows nearby. The green cover crops (bell beans, vetch, and Magnus pea) stand straight as sentries; the low eggplant beds hang heavy with late purple fruit. I harvest a few olives from the 100-year-old trees nearby, planted by the earliest farmers in the area, while the children run between the neat rows, searching for tomatoes on withered vines. Yellow-billed magpies and crows, both natives, wrestle for the best telephone pole to sit on. A red-shouldered hawk swoops by, scattering both flocks. Jeff Maurer, an ornithologist at UC Davis, believes that the birds here on Raoul's farm are much more diverse than those found on conventional farms nearby and is carrying out studies to test this theory.

Here in the Central Valley, we are starting to see an increase in the number of farms, like this one, that are friendly to wildlife. At the same time, the newest genetic tools are being used to minimize the negative ecological effects of farming. From where I stand, I can see the greenhouses holding the transgenic rice, tomatoes, and other crops studied by UC Davis scientists. Just as agriculture in the valley changed and adapted in response to population demands in the past, it is changing again in response to the need to spare our dwindling wild lands, and to preserve genetic diversity.

An excited shriek; Cliff has found three small red tomatoes. Tomatoes in December? Clearly there is some magic left in the valley. A magic created by scientists and farmers, different from the magic of the wilderness, but dependent on it as a source of new genes, clean water, and much more. Just as the fate of food production relies on stewardship of the land, the fate of wild nature is increasingly tied to the ways we farm.

Nine

Weeds, Gene Flow, and the Environment

A few minutes past sunrise on a cold, clear day in early spring, I steal out to the garden to do some weeding before my family wakes. The quiet is interrupted only by the dutiful daily refrain of the rooster and the excited chirrp! chirrp! of the sparrow. I stop by the shed to draw on gloves, pick up a hoe, and tie on my gardening belt stuffed with a small shovel and other tools. I walk past the perennial beds and underneath an arbor supporting an overgrown Banks rose to reach a sunny patch of my half-acre garden. I am particularly proud of this spot. When I moved here fifteen years ago, it was an abandoned horse pasture infested with yellow star thistle. This non-native plant with gray-green lizard-shaped woolly leaves is so toxic that if ingested by horses it causes a neurological disorder resembling Parkinson's disease in humans (DiTomaso 2006). Over the last 150 years this weed has spread over twelve million acres in California.

I don't suppose many love yellow star thistle except perhaps the bees attracted by its bright saffron-colored flowers or the honey lovers addicted to the bees' sweet product. Most find this weed a nuisance and there is no easy way to remove it. It has taken me many years of burning, mowing, hoeing and replanting with perennial grasses, to restore the old pasture. Today it is full of purple needlegrass, blue wildrye, and California poppies, native species that used to blanket the valley before human activity and invasive weeds all but obliterated them.

With my hoe I weed out a few remaining yellow star thistle seedlings, leaving the semi-restored grassland mostly clean. I survey the adjacent orchard. Already an intimidating mat of alien grasses, mostly brome and wild oats, battle with the trees—pomegranate, persimmon, grapefruit, plum, pear, peach, and mandarin orange—for open ground. Discouraged, I leave the orchard for another day and walk along the granite path past red sage and California fuchsia. In a few months, the red flowers will attract dozens of hummingbirds and the garden will be filled with the irregular drumbeat of fast moving wings. I stop and bend down to disentangle a pink-veined morning glory (also known as field bindweed) that is strangling two tulips along the path. I notice Bermuda grass creeping towards the lawn and Johnson grass invading my lavender and rosemary garden. Using a hand hoe I dig out the intruders and move on down the path to the vegetable and cut flower garden. I look around with dismay. Cheeseweed, a mallow with small and deceptively delicate pink flowers is already assaulting my strawberry and asparagus plants and curly dock is overtaking the sweet peas. Unless I do some work soon, it will seem that I am cultivating weeds rather than a crop.

My struggle with weeds over the years has made me aware of the damage they can inflict in gardens, farms, and native ecosystems. I have learned to be vigilant and untrusting of even the smallest, innocent-looking weed seedling and yank it out upon first sight. I have grown curious as to why many weeds spread invasively whereas most crops and native species do not. Both crops and weeds have pollen that can spread widely, right? So why the difference in invasiveness? To answer this question I needed to learn where weeds originate, why they persist and reproduce, and how domesticated crop plants differ from weeds. These issues have become important in the debate about the potential impact of GE crops. Some people worry that the presence of GE pollen in the environment will create a new breed of invasive out-of-control weeds that will overrun pristine environments or irrevocably alter the genetic makeup of native species.

The weeds in my garden share similar characteristics. First they are very clever. Yellow star thistle, for instance, is able to complete its life cycle quickly. It germinates with the first rains of fall, sending its roots down to depths of 6 feet or more where it sucks up all the moisture so that there is none left for the slower-germinating native species. Second, the weeds here are mostly alien and have evolved adaptations that allow them to survive and spread. For example, yellow star thistle was introduced from Europe and then to Chile hundreds of years ago. In the 1850s it traveled to California from Chile as a stowaway in alfalfa seed, where it was then inadvertently planted with the hay crop. Third, profuse production by seeds or aggressive vegetative structures is common. In midsummer, one single large yellow star thistle plant producees seedheads bearing long tan spikes that can yield 100,000 seeds. The underground stems of field bindweed can send up 1,000 new plants each season. These are some of the ways that weeds can outsmart their domesticated cousins and create constant work for gardeners.

Before I attack the cheeseweed, another deep-rooted, aggressive weed, I loosen up my old back (at least the little that remains after so many years of obsessive weeding) by bending and stretching while I think about last year's sweet corn crop. The nearly identical plants with their specially bred tassels perched gaudily on top of single stems, grew taller than me and delivered oversized ears. These traits—large fruit, reduced branching, gigantism, reduced seed dispersal, and a lack of genetic diversity—are all signs of domesticity (Ross-Ibarra et al. 2007). My corn did well- quite content with the domestic life it was bred for, where nutrients were plentiful and a slavish gardener destroyed all competition. Clearly these corn plants were not going to make it in the wild without me around.

Looking west from the garden, past the old rows of olives, I see the sunlit blue-gray foothills of the inner coast range harboring some of the wildest land in California where mountain lions and bears still occasionally surprise visitors. Between our farm and these hills are vast agricultural fields. Because of the proximity of the farms and wilderness, it would seem that the crop plants could escape to the nearby foothills, following the example of the eager weeds. They have not. On trips with my students to identify species of native plants in this area, we found an overabundance of weedy oats, bromes, and starthistle. Yet, notably absent from these weedy foothills were crops from Central Valley farms—we saw no corn, no soybean, no alfalfa, no cotton, no tomatoes, no safflower nor rice. Although these domesticated plants are also aliens—tomatoes and corn from Central and South Americas, cotton from what is now Pakistan, safflower and alfalfa from the Near and Middle East, and rice from China, any residual weediness has been eliminated through many years of breeding (Hobhouse 2005). This is one of the reasons that the genetically modified corn and cotton, grown here for 150 years, have not established in the foothills. GE cotton and corn, the primary transgenic crops grown in the Valley, are not likely to survive here either. After all, a GE crop is still a crop and crops make lousy weeds. The traits that make these plants good for farmers make it hard for them to survive in the wilderness (Berthaud and Gepts 2004).

I kneel down on the ground, pull out the remaining bindweed and cheeseweed, and then prepare the beds for planting later in the spring. The compost that we spread here yesterday is heavy and moist, a rich dark satin irregularly stained by a few pieces of eggshells—pale blue, brown, and white. I pull my shovel from my garden belt and begin digging, mixing the compost into the clay-colored soil, careful not to harm the plump slow-moving reddish earthworms. As I weed and turn the soil, I think about the concern that pollen from GE crops might drift over to the nearby foothills to create a new kind of weed that will pollute native ecosystems. What if transgenes move from a GE crop to a weedy relative? Can transgene pollen flow somehow transform a crop into a weed or change an ordinary weed into a "super" weed? Most experts say that this is unlikely. That is because it takes many genetic changes arising from a combination of gene flow and spontaneous mutations to become a weed.

There can be no gene flow, that is to say, no sex, without two willing partners. And most plants are quite choosy, preferring a close relative rather than someone outside its family. Pollen from crop plants (GE or non-GE crop) can travel around all it wants—in gusts of wind, on the pollen basket of bees, as cargo of flies or in the hands of human plant breeders—but unless the pollen alights upon a compatible mate, there will be no fertilization and therefore no seed. And if there is no progeny to pass genes onto, there can be no gene flow. In the Central Valley, genes from GE crops plants cannot be shared with the native populations nearby, because the GE crops grown here have no sexually compatible relatives in the foothills. This means that the GE species grown in this great valley are trapped. It is as if California were a large, oval-shaped, flat-bottomed platter with steep, slippery sides holding all the GE crop plants at the bottom.

Of course, as more crops are genetically engineered, the picture will not always remain so simple. This is because cultivation of other crops could potentially create problems under certain conditions. For example, most ecologists believe that if pollen carries a trait (GE or non-GE) that confers a "fitness" advantage (e.g., enhancing viable seed production), *and* it has wild relatives nearby, it could potentially establish in some environments and become invasive. If the gene confers no fitness advantage it would be lost from the population over time. If it confers an advantage and is passed on to relatives, it would be maintained in the population. For example, if a drought tolerant gene from wheat hybridizes with a related weed called jointed goatgrass and if the hybrid establishes, it could become more of a problem in the western United States.

There is certainly evidence for cross-hybridization of crops with wild relatives, but few if any of the resulting hybrids have become invasive. For example, in Quebec, Canada, domesticated GE *Brassica napus* (canola) is able to hybridize with a weedy relative called wild radish (*Brassica rapa*) (Warwick et al. 2003). According to Norm Ellstrand, a population geneticist at UC Riverside, "canola is as yet the only case known in which engineered genes from a commercial crop have been found in natural populations" (Ellstrand 2006). Although the transgenes could be found in hybrids between the two *Brassica* species, they slowly disappeared over subsequent generations and their presence, therefore, did not alter wild populations. Another study demonstrated that these Brassica hybrids actually decreased competitiveness of the wild radish species, turning this particular weed into a "wimp" (Adam 2003; Halfhill et al. 2005).

Last year, I asked Steve Strauss, a Professor of Forest Science at Oregon State University, who spends much of his time promoting public understanding of genetic engineering, about his research. He told me that all studies looking at the issues of gene flow between domesticated and wild relatives have shown that crop domestication has not benefited the wild relatives. He explained that despite intensive breeding for stress tolerance in annual crops, "there appear to be no known cases where populations that are substantially more invasive in the wild were generated as a consequence" (Strauss 2003).

Apparently is it quite difficult to turn a docile crop into a promiscuous weed.

Recently, I learned that a perennial weed with extraordinarily light pollen that cross-pollinates with at least twelve other species of grass, has been cultivated on Oregon golf courses for decades. Unlike crops that have trouble surviving off the farm, this weed, creeping bentgrass, can easily survive in the wild. Nevertheless, golfers and caretakers of fairways like this weed because it is low-growing and easy to take care of. Now, two companies, Monsanto and Scotts, have genetically engineered this weed for tolerance to the widely used herbicide glyphosate (often sold as Roundup). Other weeds on the golf course will be killed by Roundup but GE creeping bentgrass will survive in the presence of the herbicide (Watrud et al. 2004).

Not surprisingly, GE pollen behaves no differently than its non-GE counterpart. Researchers have found creeping bentgrass transgenes in the progeny of wild populations 14 kilometers away from the source field. The transgene is not expected to spread in the population in the absence of the herbicide and is therefore not a problem for gardens or the wilderness. In other words, the GE weed is still a weed, but survives no better than its non-GE counterpart. Still the case brings up interesting questions as to what should be cultivated or genetically engineered. From the point-of-view of a gardener who spends several hours a week pulling weeds and adding them to my compost pile, it makes no sense to plant weeds, let alone cultivate them. On the other hand golfers seeking smooth fairways might prefer that I take my weed-magnet of a garden elsewhere.

As I work on the garden beds, I consider another situation. Can genes flow *from* wild species *into* domesticated crops? In my garden I have so far seen no evidence for this. The native grasses I have planted occasionally reseed but are not aggressive and have not invaded my rows of vegetable crops. And because they do not share genes with other plants in my garden, I have not inadvertently created interesting hybrids. Less benign native plants can, however, create problems for farmers in other parts of the world.

Last spring, I missed a few of the best gardening days to visit the National Academy of Sciences in Washington, D.C. There, Dr. Barbara Schaal spoke about her studies with rice weeds. In collaboration with colleagues in Chiang Mai, Thailand, Schaal's research team found that a wild rice species can cross-hybridize with domesticated non-GE rice at a low frequency resulting in a hybrid variety of rice (Schaal 2007). The hybrids do not thrive in the wild because the wild rice species is better adapted and therefore quickly dominate the hybrids. For Thai farmers, however, the hybrids create a weedy nuisance. The weeds are hard to remove because the seeds shatter and build up in the soil leading to more weeds the next season. Yields of the rice crop decline proportionately with the increase in weeds.

This example, as well as others (Ellstrand 2003) demonstrate that gene flow between wild and domesticated species can occur but so far has only created problems for the domesticated *crop*, not for the environment.

⁕

While GE plants currently grown in California do not have an opportunity to interbreed with wild species, some of the California crops are exported widely and could possibly end up in environments where there are sexually compatible species. Mexico, for example, imports several million tons of corn from the United States each year.

As I continue through the beds, pulling weeds, I recall a trip I took with a friend to Oaxaca, Mexico. We traveled through rainforests rich with diversity of tree ferns, cycads, pipers, aroids, bromeliads, and orchids. We also traveled through small villages where farmers practiced subsistence agriculture. They cultivated a diversity of modern corn varieties, as well as traditional landraces—crops selected for their adaptations to specific locations and their culinary characteristics. Often landraces have been handed down from one generation to the other. These genetically improved landraces are valued because they carry prized genes for disease resistance and other agronomic or gastronomic characteristics.

I wondered if these valuable Oaxacan farmers' breeds could be endangered by pollen flow from GE corn. In a study of corn landraces in Northern Oaxaca, Ignacio Chapela, a professor in the Department of Environmental Science, Policy and Management at the University of California, Berkeley, published a paper in *Nature* providing evidence for the presence of transgenic DNA in these landraces (Quist and Chapela 2001). The published results ignited an explosion of worldwide publicity because transgenic corn had never been approved for cultivation in Mexico and there was concern that the presence of transgenes might compromise the genetic diversity of these landraces.

Although the results presented in the initial publication were widely disputed (Editor, *Nature* 2002) and then refuted by a larger peer-reviewed study (Ortiz-García et al. 2005), the paper prompted an important debate over possible biological, economic, and cultural implications of gene flow. These issues are increasingly important because Mexican corn growers want to use GE to improve productivity and poor consumers rely on this staple. Carlos Salazar, president of the National Confederation of Corn Producers in Mexico, estimates that more than 90% of small and medium growers would use GE seeds if they were available (Garret 2007). Recently, Mexico's corn growers signed an agreement with Monsanto to buy and plant genetically altered seeds.

Will a future massive planting of GE corn create a problem for local landraces? As Paul Gepts, a geneticist at UC Davis, points out, because domesticated non-GE modern hybrid varieties are now widely planted in areas of high biodiversity, "modern" genes are already present in local landraces, often introduced by local farmers who wish to generate new varieties (Ortiz-García et al. 2005). It is unlikely that a single transgene by itself would reduce the genetic diversity of native populations to a greater extent than is already occurring.

Fortunately, at the policy level, native landraces have actually benefited from the discussions on GE corn in Mexico. The GE corn debate has led to greater recognition of the value of indigenous landraces and Mexican growers plan to initiate activities to protect these landraces, including setting up a maize germplasm bank. They recognize that cultivation of modern crops, GE or non-GE, needs to be examined carefully in order to safeguard the center of genetic diversity where pollen flow could impact the genetics of local plant populations.

Across the street from my garden and our farm is a large conventional farming operation. The grower plants a rotation of wheat, alfalfa, sunflowers, and corn with an occasional watermelon seed crop. Last year the field was planted with yellow corn for animal feed. I don't know if it was GE corn but it may have been. In any case, Raoul and I are not concerned about gene flow from these plants. First, over the fifteen years I have lived here, growing white sweet corn nearly every year, I have never noticed any yellow, hard kernels on my sweet corn ears. If I had found even one yellow kernel it would have been a sure sign that pollen from across the street or elsewhere had pollinated our plants. Although viable corn pollen grains have been found more than half a mile (800 m) from their source (Ellstrand 2006), 98% of the pollen remains within a 25–50 meter radius of the corn field (Sears and Stanley-Horn 2000; Pleasants et al. 2001). So every year we continue to enjoy our tender white corn while our neighbor feeds the yellow hard kernels to his cows.

Even if some pollen had made it this far, a transgene or two would certainly not cause any harm to our garden, farm or our health. Virtually all leading scientific panels that have convened on this manner (National Research Council 2004; GM Science Review 2003) have agreed that pollen drift from approved GE varieties in the United States does not pose any conceivable increased health or environmental risk. We are also familiar with the USDA National Organic Program (NOP) standards and know that produce from our farm cannot be decertified if genetically engineered pollen inadvertently mixes with our crop. To date, no grower has ever lost certification due to the presence of a transgene in an organic product (Hawks 2004).

Despite our views, we realize that not all growers are comfortable with pollen flow from GE plants. This is mostly because some consumers of organic food want it to be "GE-free." Although a large majority of farmers (92%) surveyed in the United States report no direct costs or damages related to the presence of GE in their crops, 2% indicate that they have lost sales due to the perceived risk of transgenes (Walz 2003). There is even some evidence that buyers have begun requesting that organic farm products be tested for the presence of transgenes.

In response, some farmers have paid for a highly sensitive genetic test called the polymerase chain reaction (PCR) to detect transgenes. Using PCR, minute quantities of transgenic DNA can be detected, even in a truckload of corn (see table 4.1). Curiously, while some consumers oppose even trace amounts of transgenes, they have readily accepted other kinds of unintended biological material (e.g. insects or rodent feces that can also be readily detected by PCR) that may have been unintentionally mixed with the food product while processing. They even accept a small amount of pesticide drift on organic crops, even though certain pesticide applications *do* pose increased health risks. In fact, despite these risks, we can sell our produce as certified organic even if a limited amount of pesticide drifts onto our fields. This is because the U.S. organic industry recognizes that some level of pesticide drift from conventional agriculture is inevitable, and the NOP standards allow for the marketing of certified organic products containing some pesticide residue (less than 5% of the EPA allowable pesticide residue). Unlike for pesticides, current regulations in the United States do not specify an acceptable threshold level for the presence of transgenes in an organic product. As long as an organic grower takes precautions to mitigate gene flow between fields they can sell the products as certified organic. The European Union has taken a different approach. In 2007, the EU Agricultural Council adopted a law that allows food containing up to 0.9% GE material—acquired through accidental or unavoidable crosspollination—to retain a label of "EU organic" (Adam 2006). This law provides some assurance to organic growers that they can continue to sell their products even if trace amounts of transgenes are detected. Despite this obvious benefit to growers, some U.S. businesses and governmental organizations oppose passing a similar law in the U.S, believing that regulatory thresholds for biological material should be science-based. Because trace amounts of known transgenes have no known harmful health or safety effects, any threshold is somewhat arbitrary.

I have completed weeding the beds and fill my hands with the satiny soil that offers itself up to my daydreams. As soon as the risk of frost is past, I will plant a collection of sunflowers. I imagine velvet petals with colors of yellow cream, pale gold or deep burgundy surrounding showy floral centers painted dark chocolate, yellow, or light green. Masses of bright red and gold bicolor blooms will crowd lopsided summer rows. I will harvest the exquisite flowers and set them on our summer table accompanied by basil-flavored pasta, homemade amaranth sesame seed baguettes, and perhaps a plum tart. Some flowers I will leave in the garden so that the birds can enjoy the ripe seed.

As I stuff my tools back into my gardening belt and get up off my knees in the warming air, I marvel that it is possible to fear these flowers. But even sunflowers have not escaped controversy.

Twenty years ago organic farmers in this area began growing specialty sunflowers to sell for cut flowers. Although most of the pollen from organic sunflowers does not travel further than 3 meters, some of it can travel up to distances of 1000 meters, which can cause problems for growers of certified sunflower seed (Aria and Rieseberg 1994). If stray organic pollen should land on a sunflower grown for seed and hybridize with it, the resulting seed will no longer be purebred, reducing the value of the crop. This is the reason that sunflower seed growers in the valley were concerned about gene flow from organic sunflowers.

The certified seed growers and organic flower growers worked out a comfortable arrangement. The seed growers gave the organic growers seed that produced pollen-free flowers. This allowed the organic growers to continue to sell the flowers and elimi-nated the risk of gene flow. This compromise offers a good example of how discussions among neighbors can lead to mutual benefits. Because California farmers grow 350 recognized crop and livestock commodities under a variety of farming conditions, often on adjoining fields, good communication and common sense is key to dealing with pollen flow (see box 9.1). These principles apply to all crops—GE, organic, or conventional. Unfortunately, not all stories end so happily as the sunflowers.

BOX 9.1 **Coexistence**

One model for coexistence between GE and non-GE crops is the program established for publicly owned land in Boulder County, Colorado (Byrne and Fromherz, 2003). The county leases about 4,000 acres of cropland to farmers, some of whom have chosen to grown insect resistant or herbicide tolerant GE corn. An advisory committee of farmers, scientists, and concerned citizens developed a set of protocols to minimize cross-pollination to nearby non-GE corn fields. The protocols include grower notification to the county of their planting intentions, communication among neighboring farmers to work out an acceptable coexistence plan, and establishment of a 150-foot buffer zone between fields to minimize cross-pollination.

According to Pat Byrne, a member of the advisory committee and association professor at Colorado State University, the size of the buffer zone was determined from multi-year cross-pollination studies in Boulder County, showing that 150 feet was sufficient to keep cross-pollination below 1%. "We used the blue kernel trait to track cross-pollination from a central field of blue corn to a surrounding field of yellow corn," explained Byrne. "It felt good to apply my esoteric knowledge of kernel pigments to a societal issue like coexistence." He continued, "A guiding principle of the protocols was shared responsibility for preventing unwanted cross-pollination. The county and GE crop grower are required to provide a sufficient buffer to keep cross-pollination below 1%, and if the non-GE grower requires levels below 1%, it is the responsibility of that grower to provide the extra measures."

Sally Fox is an unlikely player in the cotton business, an industry dominated by male farmers, who increasingly plant GE cotton. You can see this in her appearance (loose cotton dress, bright blue eyes, friendly smile, unreserved enthusiasm, and graying hair) and in the crop that she grows—certified organic colored cotton.

Sally knows a lot about contamination. Educated as an entomologist and with a background in cotton breeding, she began her first cotton business in 1986 in the San Joaquin Valley in California. A few years later she sold her first crop to a Japanese mill. It wasn't long before her business took off. Levi's, L. L. Bean, Land's End, and Esprit all became customers of her "natural" cotton. Soon Sally was running a $10 million business. And soon there were problems.

Sally's neighboring cotton growers were afraid that Sally's organically grown, colored cotton would contaminate the white cotton crops grown in the same valley and processed in the same gins. They imposed strict rules on her operation, which forced her to move to Arizona in 1993. Six years later, Arizona cotton growers did the same thing, and Fox had to relocate again, this time to the Capay Valley near Davis. She is now mostly retired; devoted to spending time raising her five-year-old daughter.

"The contamination fears were overblown." she told me over brunch at a neighboring farm a few weeks ago. "Although I did learn that mixing can actually occur—there may be someone who will break seed segregation rules if they think it will save time or increase their profits—a bit of colored cotton would not have destroyed the industry. It is not hard to pull out plants producing reddish brown cotton from a field of white before they cause problems. At that time however, organic was not a big business and it was relatively easy to pass laws to exclude my operation."

In California there are no laws governing pollen flow from GE or other crops, growers need to talk with each other to avoid a "Sally" situation where a well-meaning, innovative, and productive grower will be forced to leave the county.

Mary Bianchi, an energetic Horticulture Advisor for the University of California Cooperative Extension, one of the organizations working to improve dialogue, recently told me that the idea is to move past the discussions of GE as being "good" or "bad" and to make the system workable for both the California organic industry and conventional growers. Achieving 100% purity for any agricultural product is extremely difficult but through separating fields spatially, staggering planting dates, or growing varieties with different maturity dates farmers have minimized or eliminated cross-pollination. Segregation of varieties during harvesting, shipping, and processing also helps prevent the inadvertent intermingling of organic and conventional produce.

In the waxing morning light I walk back to the shed to store my tools, wiping the soil from my hands onto my apron. I stand empty handed in the silence, the sun warming my bones, and look across the street at the neighbor's field of alfalfa and wonder if I should strike up a dialogue. I am not worried about errant alfalfa pollen but I dislike noise and pesticide drift over our garden, home, and school. I know that next month, probably at dawn on a clear, still morning like this, the drone of planes will disrupt the calm as it does every spring that alfalfa is planted. The planes will spray the fields with a pyrethroid pesticide called Warrior meant to halt the spread of alfalfa weevils. If someday there is a GE or organic approach to combating these pests, I would gladly embrace it. I would prefer some benign GE or organic pollen blowing around rather than the almost-unbearable noise and the sickly-sweet odor of pesticides. As I mount the stairs of the front porch, I wonder if I can convince my neighbor to consider some new ideas.

Ownership

Ten

Who Owns the Seed?

I finish work at the farm at 1:30 P.M., pick a few vegetables, hop in my car, and head home for lunch. As I turn into our gravel driveway, the car tires crackle and my stomach growls. I'm starving, not having eaten anything since breakfast. I gather my tea mug, a bag of kale and tomatoes, and my sweater and head to the mailbox, my arms full.

I check the mail every day with a sense of anticipation, hoping for checks and invitations, but receiving bills, catalogs, and PennySaver mailers instead. The selection today is mostly junk: a credit card offer, local coupon book, catalogs and, wait a second—jackpot! Here is something even better than checks or invitations: the 2006 Johnny's Selected Seeds catalog.

Stuffing the mail under my arm, I head into the house, and dump everything on the island in the kitchen. Moments later I find myself sitting in a comfortable chair in the living room with the Johnny's seed catalog. All thoughts of food have vanished, and I'm not aware of how I got here. My brain has shifted all its attention to the seeds, the plants, and their traits.

Johnny's Selected Seeds is where I have bought the majority of vegetable seeds for my various organic farming operations. I like Johnny's. The owners cater to organic vegetable farms of all sizes. They sell a lot of seed from other companies, and also have developed some of their own varieties. The company's focus seems to be on the more innovative varieties that are early, uniform, and disease resistant. Years of plant breeding research go into developing a new variety. I see from the prices of the hottest varieties that the seed companies that developed them are trying to make back their investment, and then some. This seems fair in the sense that companies need to make a profit to stay in business, and I wonder how they protect their varieties from being copied by other companies or farmers or seed savers. I also wonder if I can afford to buy the seed.

Johnny's owners, Rob Johnston and Janika Eckert, are featured on the catalog cover holding a basket full of long, red peppers. It's unusual to picture the owners on the cover of a seed catalog, because the vegetables are the stars of this book. Perhaps this is an effort to show how these two are accessible and proud of their product, but it also looks like they are there to protect their latest variety from unsavory characters waiting to steal their seed.

As I browse methodically through the catalog, I can't help noticing that most of the new varieties are hybrids—pricey hybrids. A hybrid is the offspring from plants of the same species but of different varieties; the resulting offspring carry half the genes from each parent. Sounds simple, but the process of hybridization takes time and effort. First, a breeder starts by creating two "inbred" parent lines, over many years. To do this the breeder allows each plant to "self-pollinate" for many generations until they attain genetic uniformity and will not segregate for new traits in the next generation. Then the breeder cross-pollinates plants from these inbred lines by placing the pollen (the male gamete) from the parent of one line onto the pistil (carrying the female gamete) of another parent of the other line (see box 4.1). For some reason that is not entirely clear, in some plants the cross-pollination of in-bred parents results in offspring with "hybrid vigor," which typically means higher yield. Unfortunately, if the farmer replants seeds that the hybrid itself produces, the plants that grow from these seeds are not the same as the parent—they do not "breed true." Instead the off-spring include a varied assortment of types because each of the new seedlings inherits an unpredictable mix of genes from the hybrid parent. From the seed company's point of view, this is great, for each year the hybrid seeds have to be created anew by the seed company. They are expensive, but most organic growers buy them, because the hybrid vigor, uniformity, disease resistance, yield, and sometimes taste, are deemed to be worth the extra cost. And, most farmers are unwilling to create their own inbred lines by cross-pollination each year. Few have the time to be both a breeder and a farmer. In any case, Johnny's sells many wonderful hybrids: Packman broccoli, Nelson carrots, Ambrosia melons, Big Beef tomatoes. These are just my favorites—the list goes on and on.

The first documented, intentional hybrid was created in field corn by G. H. Shull in 1909 at Cold Spring Harbor, New York. I don't know if Mr. Shull realized what he had done for the seed industry. From his writings, it seemed he knew making hybrid seed would be more expensive because it took more time, but he wasn't sure if the increase in yield would cover the extra cost of the seed (Shull, G. H. 1909). It took a while for the idea of hybridization to gain popularity. At first it was viewed as imprac-tical and too complex, and farmers resented having to buy new seed each year. In 1930, only 1% of the corn crop was hybrid. After several years of drought, however, when hybrids responded better than the traditional varieties, their use rapidly increased. By 1940, 30% of U.S. corn was hybrid. By 1970, 96% of the U.S. corn crop was hybrid (Federoff and Brown, 2004). Today farmers now can buy hybrid seed for popular veg-etable crops like tomatoes, broccoli, melons, peppers, and sweet corn. With the ascent of hybrids, seed companies now control the supply of the most widely used varieties and the seed is much more expensive albeit commensurately higher yielding than other types of seed (see box 4.1). In 1920, corn yields were approximately 20 bushels/ acre. Today growers of hybrid corn harvest about 160 bushels/acre. In corn-growing competitions, up to 300 bushels/acre have been achieved (NCGA 2004).

Many of the seed companies producing hybrids today are large corporations. Similar to the trend in the organic food industry, corporations have been buying seed companies. In January 2005, Monsanto bought Seminis, which had previously purchased Peto Seed and Asgrow Seed. Now Monsanto competes for a large segment of the U.S. vegetable seed market. The company that developed GE corn, cotton, and soybeans is now the company that controls many of the hybrid vegetable varieties organic growers like to grow.

Not all the varieties in the catalog are hybrids. When a parent plant is fertilized by another plant of the same genetically stable population it's called open pollination (OP). The offspring of these parents have traits that very closely resemble the parents, and seed can be saved from one generation to the next. Before the invention of hybrids, farmers planted OP varieties, selected the best and saved seed from these to plant the next season. Through this selection a farmer could direct the evolution of plants to his or her ends. For example, with tomatoes, which are naturally self-pollinating, with a low percentage of out-crossing, the farmer plants a variety and chooses the largest, crack-free, tastiest ones, and saves the seeds. As the farmer continues to select for these traits, the gene mix of the tomatoes become slightly more uniform each year, and after many generations, the tomatoes may become a little larger, have fewer cracks, and taste better, but the improvements are limited. If a particular variety only has genes to produce a five-ounce sized fruit, the tomato is not going to get much bigger than that unless there is a genetic variant somewhere in the population.

Plant breeders trying to improve a particular OP variety will cross-fertilize it with other varieties within the same species that may have useful traits. After the cross is made and the plant produces seed, the breeder plants the seed and then selects for plants that contain the desired trait. The breeder will then try to stabilize the selection so it breeds true in succeeding generations. The process takes years. At the student farm we are presently part of the Organic Seed Partnership (OSP), funded by the USDA. The group's goal is to develop vegetable varieties well adapted to organic production. Some of the varieties we are testing on the farm are ones that University plant breeders, particularly Molly Jahn, professor of plant breeding from Cornell, have created through crosses. When I asked Matt Falise, a vegetable breeder in the Department of Plant Breeding and Genetics at Cornell, who helps organize the OSP, how long it takes to develop a new OP variety, he estimated about eight years. Molly estimates it could be between three and thirty years.

Anyone can save seeds from OP plants and many companies, organizations, and home gardeners do just that. Groups like Seed Savers Exchange have specialized in saving OP varieties that may have been passed down by somebody's grandmother, or may have been discontinued by a seed company because another variety was developed that was considered an improvement. Many of these older varieties are called heirlooms.

I continue browsing through the catalog and get stuck on the heirloom tomato page. The heirlooms most commonly grown around here are tomatoes. Johnny's offers quite a few, including Brandywine, Striped German, Cherokee Purple, and Pruden's Purple. These heirlooms generally taste better or are more exotic looking than the hybrid red slicers, but they soften easily, are lower yielding, crack readily, and are susceptible to many diseases. Local organic growers like to grow the heirlooms though, because they sell from $20 to $30 for a *ten*-pound box, compared to $15 to $25 for *twenty* pounds of hybrid, red slicers.

The most popular heirlooms, such as the Brandywine tomato (which many consider the best tasting), are offered by almost all the seed catalogs I have. The Territorial Seed Company catalog understates a not-so-endearing trait, "Not a heavy yielding tomato," which probably explains why for many years it was not commercially available. Although a fair amount of Brandywine seed is once again being sold, it is probably not as profitable for the seed grower or the seed company. Johnny's sells it for $15.90 for a quarter ounce, about 2,187 seeds. Compare this to my favorite, the high yielding, crack-free hybrid, Big Beef. Johnny's is selling the same amount of seed for $169.49. It does take more work to produce the hybrid, but is it really ten times more?

If you want some perspective on heirloom vegetable varieties, find a reprint of *The Vegetable Garden* by M. M. Vilmorin-Andrieux, 1885. The book has illustrations, descriptions, and growing practices of garden vegetables of France in 1885. It provides a baseline, of sorts, with which to compare today's vegetables and those of 125 years ago. Vilmorin states that for broccoli, "instead of producing a head the same year in which the plants are sown, it usually does not do so until early in the following spring." Modern broccolis have certainly come a long way since then, with some varieties producing heads within 60 days. On top of that, the broccolis described in Vilmorin's book are white-headed instead of green! Carrots at the time were sometimes orange, but more often red, yellow, or white. Some of the heirloom varieties we use at the student farm are listed in the book, like Early Nantes carrots, Egyptian beets, and Jersey Wakefield cabbages. However, most of the varieties I have never heard of, and some of the vegetables themselves seem like they are from a different planet. I wonder where all the genes have gone that coded for these different colors and shapes. Steve Tanksley and Susan McCouch, geneticists at Cornell University, estimate that modern tomato and rice varieties contain only a very small percentage (1–5%) of all the possible traits available in their wild relatives because over the years many of these traits were selected against through domestication and breeding (Tanksley and McCouch 1997). I imagine that this has happened to virtually all other improved vegetables as well.

Reading Vilmorin's book I get the sense that human beings are driven to breed plants, improve plants, and come up with something new and better. Ironically, this has meant that diversity is reduced because traditional breeding techniques select for a few important traits and discard the plants not exhibiting them. Potentially useful genes that encode for traits that cannot easily be seen, tasted, or smelled are lost. Such lost genes can be recovered only by going back to the wild ancestors of our crop species and landraces that have been conserved by traditional farmers throughout the world. This is quite difficult to do without help from modern genetic techniques.

Are the beautiful and tasty heirloom varieties protected and owned exclusively by a particular company? No. Organic seed companies, like Seeds of Change, have programs to improve the quality of heirloom varieties by growing many individuals of a particular variety and selecting for individuals that exhibit the best traits. We have done several variety trials for Seeds of Change at the student farm, and I have had a chance to grow and taste many wonderful OP varieties that are as good as or better than hybrids—Imperial eggplant, Crimson Sweet watermelon, Orange CA Wonder pepper, Kurota carrots, Early Green broccoli and Viroflay spinach are all very satisfying to grow and eat. But there is no mechanism for preventing growers, companies, or home gardeners from reproducing and saving (and selling) the seed. For a newly created OP variety (not an heirloom), the situation is different.

As I leaf through the pages of the Johnny's catalog I notice another icon used with open-pollinated varieties: PVP. Checking the Key to Vegetable Symbols, I see this is defined as "Plant Variety Protection—unauthorized marketing of seeds prohibited." The PVP Act was enacted in December 1970, to provide legal, intellectual property rights (legal entitlements) protection to developers of new open-pollinated varieties that are propagated by seed (Plant Variety Protection Office 2007). The act was toughened in 1994 to prohibit the sale of farm-saved seed without permission of the variety's owners, and the length of protection was extended to twenty years. Under the PVP Act, farmers may save seed of PVP varieties for use on their own fields but can't sell them. The purpose of PVP is to encourage the development of new non-hybrid varieties by allowing breeders to recoup money spent on development. I looked at the PVP website (Grin 2006), which lists all of the protected OP varieties, and was amazed by the number and diversity of plant varieties there. While there is some debate over the effectiveness of the PVP act in protection of new OP varieties, there is no doubt that seed companies think that it is better than nothing. But it is not cheap to register a variety. In 2005, the cost was $5,150, enough to keep the backyard gardener out of the variety protection business. In an e-mail, Johnny's owner, Rob Johnston explains the value of PVP:

> We have PVP on several of our own varieties, and we sell many more PVP'd varieties bred by (and PVP'd by) others, while PVP still allows farmers or gardeners to save seeds for their own use PVP disallows the variety to be used as a parent in a hybrid,

and disallows its unauthorized production and marketing. A PVP label acts as a kind of no-trespassing sign, and potential pirates usually avoid the variety. However, if there is a violation, the holder of the PVP has to do the prosecuting. We've never had to pursue anyone.

Hybrids are inherently protected by the fact that the originator maintains the parents, and, thus, has a monopoly on the seed supply. Some companies, however, will PVP parents of hybrids, which would prevent one or both from being stolen and used. For the record, I prefer the "respect" method of protecting intellectual property to the legal method, e.g. PVP. If we find that some seed company has stolen one of our varieties, I like to think that I could call them and get them to stop. (R. Johnston, personal communication, 2006)

I am about three-fourths of the way through the catalog, dazzled by the pictures of plump hybrids and beautiful open-pollinated vegetable varieties, but when I start adding up the bill for my choices, I get into triple digits very quickly, and I start to think that maybe we should just grow and save our own seed at the student farm.

Over the past ten years we have saved seed from basil, tomatoes, parsley, chard, Stutz supreme melon, arugula, cilantro, onions, watermelons, garlic, and potatoes. In the educational sense, it's fine to save seed. To see your favorite vegetable mature, flower, and make seed is experiential learning at its best. In the farming sense, however, saving seed is often a pain in the neck.

Last year at the farm, my students and I decided to save arugula seed. In order to get seed from arugula (one of the easies crops) we needed to leave it in the ground for a couple of months longer than we would have if we had just harvested it for greens. More months in the ground meant more irrigating and weeding, when the bed space being used by the arugula could have been planted with something else. Then when the arugula went to seed, it produced a lot. We didn't have a combine or a mechanical seed harvester, so we harvested the seed by hand. In the case of arugula this meant stripping off dried pods of seeds and putting them into a bag. A fair amount of seed was lost as the pods broke in our hands and fell to the ground. Eventually, after a couple of hours we ended up with a few pounds of seed mixed with quite a bit of chaff. We were lucky enough at the student farm to have a simple mechanical seed winnower that more or less separated the seed from the chaff. After another couple of hours of cleaning, we ended up with less weight than we started with, but much cleaner seed. It took a couple of our students four or five hours to harvest and clean a pound of seed. Johnny's sells a pound of organic arugula seed for $26.15, but even adding tax and shipping meant that we were working for about $4/hr. (This doesn't include the cost of growing the crop either. Hopefully we covered those costs in the arugula we harvested and sold. But when a farmer is growing the crop just for seed, everything must be done very efficiently for it to be profitable.)

I recently asked Paul Holmes, a partner in Terra Firma Farms, a very successful organic farm in Winters, CA, if he saved any seed this year. Terra Firma grosses close to $1 million a year selling through a large CSA (community supported agriculture, which is a subscription produce service), farmer's markets, and retail and wholesale outlets. Paul said he wanted to save some of the orange, heirloom tomatoes called Valencia, for which he was having a hard time finding seed, but had never gotten around to it. He, and everyone else on the farm, was busy too busy to save seed, which is typical of organic farms in this part of California.

In other parts of the country like New England, New York, the Northwest, and the Midwest (well, maybe everywhere but California), saving seed is much more common. At a recent meeting of Organic Seed Partnership participants (where I was the only grower from California) I was amazed at the extent of farmer participation in on-farm variety trials and of seed saving through out the country. Part of the explanation may be that California growers have higher land costs, and therefore can't afford the field time needed to save seed. Or, somehow growing several crops, year-round in an agricultural paradise, makes one so busy, there isn't time to save seed, or even reflect much on what's important. Another reason might be that in the New York/ New England area, the OSP has a mobile seed-cleaning trailer that goes from farm to farm to facilitate seed cleaning by local growers. Perhaps if this technology were available here, more growers would save seed.

The ability of growers to save seed does help to keep seed companies from getting rich selling OP varieties. If OP prices get too high, growers have an incentive to save seed. At reasonable prices it is easier to let the seed companies provide the seed. In addition, they generally do a better job of maintaining seed purity and quality. If hybrid prices get too high, growers can switch to OPs instead, and save seeds. This can be a difficult choice if a specific trait like disease resistance, size, or uniformity is needed. Yields may also be less.

Reading about heirloom tomatoes reminds me that I'm still hungry. I'd like to eat a sandwich with Brandywine tomato slices, but it's winter, so instead I settle for a couple of quesadillas with salsa and canned heirloom, organic Jacob's cattle beans. I sit down to eat, catalog once again in hand, dripping salsa on the pages.

Although Johnny's caters to organic growers that doesn't mean Johnny's sells only organic seed. The USDA NOP standards state that organic growers must use organically grown seed if it is commercially available. If not, then growers can use conventionally grown seed that has not been treated with any prohibited materials like fungicides. Johnny's sells some organic seed, more every year since 2001, in fact, but many of the varieties they sell are hybrids, and the majority of hybrids are not organically grown. Until the last couple of years, there were no organic hybrids, so if growers were choosing hybrids, they were buying non-organic seed. Recently, Johnny's has offered more organic hybrids, like Red Ace beets, and three hybrid sweet corn varieties, and I think that organic hybrid seed will become much more common in the next five years. Johnny's does offer an increasingly long list of certified organic, OP varieties as well, with many choices of lettuce, tomatoes, cucumbers, and greens available.

I continue to fill out the order form; my mind filled with all the information and intrigue that lies between the lines of the seed catalog. I realize that I value the qualities of hybrids: the higher yield, disease resistance, uniformity, and in some cases (like Nelson carrots), the taste. Even though the hybrid varieties are well protected and dearly priced by their developers, I seem, once again, to be willing to cough up the money to pay for the traits I value. If the prices get too high, I will shift to OP varieties. If I get totally fed up with seed prices, I can go back to seed saving.

Oh, one last thing about Johnny's Selected Seeds that I didn't mention. None of Johnny's seed is genetically engineered. In fact, in the beginning of the catalog there is a statement indicating that they are proud to be a member of Safe Seed Initiative, and pledging that they do "not knowingly buy or sell genetically engineered seeds or plants." And they provide this explanation:

> The mechanical transfer of genetic material outside of natural reproductive methods and between genera, families, or kingdoms, poses great biological risks, as well as economic, political, and cultural threats. We feel that genetically engineered varieties have been insufficiently tested prior to public release. More research and testing is necessary to further assess the potential risks of genetically engineered seeds. (Johnny's Selected Seeds 2006)

As I read this, two thoughts come to mind. First, it strikes me as odd that the Safe Seed Initiative is concerned about GE varieties, but not varieties grown using pesticides, because the use of pesticides is an on-going problem. In California alone, there were 828 confirmed pesticide injuries in 2004 (DPR 2006). As far as I can tell, there were no reported injuries due to GE varieties in California, the United States or the world. While GE varieties for herbicide-resistant crops or crops containing Bt have other issues for organic farmers (see box 5.3), and would not have been my first choices as crops to engineer, they haven't physically injured anyone since they were first planted in 1996 (NAS 2004). They also haven't escaped into the wild or created super-weeds, while pesticides are still contaminating rivers and underground aquifers. If the Safe Seed Initiative is concerned about biological risk, then why aren't they concerned more about pesticides and, therefore, advocate selling only organically grown seed? Presently, there are only two commercially available GE vegetable species. Asgrow Vegetable Seeds (now owned by Monsanto) has a few yellow summer squash and zucchini varieties (same species) that are resistant to zucchini yellow mosaic virus, watermelon mottle virus, and cucumber mosaic virus. Syngenta markets a GE sweet corn that has a Bt gene to control corn earworm and European corn borer (see figure 5.1).

Because there are only two, Johnny's and other seed companies aren't giving up much by avoiding GE varieties. On the other hand, if Johnny's were to drop all the varieties grown with pesticides, (i.e., most of the hybrids and a good share of the OPs) many varieties would be unavailable.

Second, I notice that the Safe Seed Initiative has clumped all GE varieties together and hasn't analyzed each one on a case-by-case basis. To me this is throwing the baby out with the bath water. GE plants approved by EPA or FDA have risks of unintended consequences ranging from extremely low to low, and also have a range of benefits, including some that would fit well into our criteria for a sustainable agriculture (box P.3).

It seems to me that the Safe Seed Initiative GE policy slows the development of varieties that could facilitate ecological farming. What if a tomato plant is genetically engineered with another tomato gene? That's the same sort of genetic transfer that occurs with open-pollinated plants in nature or could be done by plant breeders using traditional methods. The advantage of GE instead of traditional plant breeding would be that only one gene would be introduced—the gene that expresses the desired trait and it could be done in less time. If, for example, you wanted a Brandywine tomato to be resistant to nematodes, you could put the nematode-resistant gene from Red Sun tomato (also sold by Johnny's) into Brandywine. With the addition of only one gene, the Heirloom Brandywine would retain all of its tastiness. These tomatoes would seemingly not pose any negative economic, political, cultural, ecological, or health threats. And if there were other tomato genes that could be put into Brandywine that would increase the yield, make it resistant to diseases, eliminate cracking, and make it just a little firmer, you'd have a heck of tomato. At such a future time, would Johnny's sell the seed? And would organic growers grow it and consumers eat it? That may depend on who owns the genes.

Eleven

WHO OWNS THE GENES?

One evening, in a ceremony at the annual meeting of the American Society of Plant Biologists, in the Seattle Convention Center's cavernous ballroom, Richard Jefferson, the chairman of the Center for the Application of Molecular Biology to International Agriculture (CAMBIA), a nonprofit research institute, stands next to the podium. A casually dressed man in his early fifties, he listens as the ASPB president, Roger Hangarter, thanks him for his outstanding contributions to science and humanity and awards him the ASPB Leadership in Science Public Service Award.

"Jefferson's contribution of tools for genetic engineering could have a major impact in making these technologies freely available in the United States and developing world. His contributions are scientific but with substantial public service implications," Roger explains. "Richard has succeeded in working through a complex web of patents to make plant transformation technology more widely available."

When Roger finishes, Richard steps to the podium to begin his talk. He speaks directly to the audience without relying on any slides. "People lose interest when the speaker drones on and on, pointing at slides. It is better to lose the props," Richard told me earlier in the evening. Most of us here are not as eloquent as Richard or as comfortable winging a speech in front of an international crowd of hundreds of plant biologists. Richard, however, is a well-known maverick in the plant biologist community in the sense that he does things his way, on his own, and is usually successful. He has our full attention.

I first met Richard late one evening in 1992 at a meeting in Bali on rice biotechnology sponsored by the Rockefeller Foundation. Over a period of fifteen years, The Rockefeller Foundation spent $100 million to foster cutting-edge genetics research aimed at helping rice farmers in the developing world. The goal of that particular tropical gathering was to bring the world community of rice researchers together to discuss scientific results. It was a warm, humid night, and Richard was playing guitar on the steps of his small cottage. He is a dedicated musician, composing and performing on both the guitar and the mandolin. As I walked by, he leapt up, introduced himself, and then proceeded to talk almost nonstop. I quickly discovered two things: that Richard is delightfully approachable—within minutes of knowing him he will tell you about his love life, his latest scientific success or failure, and how he is feeling about a particular experiment—and that he could manage to be melancholy and passionate about a subject at the same time. That night he talked excitedly about the Rockefeller Foundation's goals to make the tools of biotechnology free and available to researchers in less developed countries. "Why should scientists in poorer countries be dependent on the developed world for needed biotechnology advances?" he asked rhetorically.

That year Richard was busily directing programs in China, Vietnam, and Africa to introduce scientists to modern genetic techniques, fierce in his determination that not only those in the developed world should retain the knowledge and assume the benefits. Yet he was deeply worried that the grant monies he was using to support himself would soon dry up because of "the inertia of the status quo that determines research strategies." Then he would have to support himself in a more conventional way, as a professor, for example.

Many years later, Richard is still successful in raising funds to support the singular path he has chosen and is still helping to make the tools of biotechnology widely available. Today, at the conference, he describes his view that the current system of patent ownership is stifling innovation and giving plant biology a bad name. He tells us that for years, he traveled constantly. "I used to be able to walk into a bar anywhere in a world and get a free beer. I would sit down, look around and start talking with someone. Before too long they would ask, 'What do you do for a living?' I would answer, 'I am a plant genetic engineer.' 'Cool!' They would say, 'that means you create new plant varieties to help feed people. Let me buy you a beer.' Now they look at me, look away and buy someone else a beer. And why is that? It is because today people associate plant genetic engineering with big chemical companies and restrictive patents."

He has a point; recent data indicate that although about one fourth of the patented inventions in agricultural biotechnology were made by public sector researchers (e.g., public universities), many of these inventions are exclusively licensed to private companies (Atkinson et al. 2003). Even more problematic, the private sector is becoming greatly centralized through mergers and acquisitions into a global oligopoly dominated by five firms that are also major marketers of pesticides (Monsanto, Dupont-Pioneer, Syngenta, Bayer, BASF). These mergers were made in part to accumulate the intellectual property (patented technologies and genes) portfolios necessary to produce GE crops and in part to gain control over a new technology that is threatening their pesticide markets (Toenniessen 2006). Monsanto now makes more money from seed sales than from glyphosate (Roundup) and the margin is growing rapidly. What this means is that the private companies now have even more control over who uses the technology of genetic engineering. If a particular aspect of the technology is key to the entire process, say for example, the means to introduce a gene into a plant, denial of access to a single technological component is essentially equivalent to denial of access to the entire process. This "exclusive licensing" by universities of key aspects of GE technology to private corporations greatly restricts the ability of the public research sector to develop new crops using GE. Furthermore, with declining financial support for public sector research and development much of the development of GE has been left to the private sector (AAAS analysis of President's FY05 budget projections [May 2004]). The result is limited development of new crops, particularly of subsistence (e.g., cassava) and specialty crops (e.g., strawberries, apples, lettuce)—the historically important work of public-sector agricultural research. Furthermore as the cost of regulation increases, the public sector increasingly cannot afford to commercialize transgenic crops (box 11.1).

Although the concept of "public good" is a key mission of university scientists, many are concerned that this mission will be blocked if the most valuable technology is exclusively licensed to or owned by a few major companies that increasingly control seed production, as well as herbicide and pesticide products. Similarly, many organic farmers and other small growers are afraid that a blizzard of patents on GE tools will give these companies even more control over agriculture.

BOX 11.1 **Regulatory Costs**

At the time of development of GE papaya (see box 4.2), regulatory costs were so low that a university professor such as Dennis Gonsalves could pay for the costs from a small grant. Today regulatory costs are quite high, effectively excluding nonprofit groups from bringing crops to market.

The current governmental regulatory regimes for GE were created largely because of potential biosafety issues concerning genes imported from distant species (e.g., bacterial Bt). Now, however, they are applied to genes whose source and effects resemble those of traditional breeding (e.g., the rice *Xa21* gene). This indiscriminate regulation imposes large costs (e.g., $50,000 to $50,000,000 per new variety) that impede the delivery of public benefits from genomic research. For example, if a university research lab genetically engineers a Brandywine tomato for resistance to nematodes, few organizations would be willing to pay the associated regulatory costs needed to bring the new variety to market. Universities and small growers could certainly not afford these costs. And large seed or biotechnology companies, who have the funds, may not be willing because of the small market. These are some of the reasons that a few scientists and policy makers advocate regulations that distinguish between classes of GE plants. For example, high confinement would be required for highly toxic or allergenic pharmaceuticals and proteins, whereas low confinement would be used for genes that modify metabolism in a manner similar to that of natural or induced mutations. At Steven Strauss, a professor in the Department of Forest Science at Oregon State University explains, "If regulatory costs and hurdles were significantly reduced, it might promote GE crop development by small

(continued)

BOX 11.1 *Continued*

companies and public sector investigators. Given the widespread suspicion of the power and ethics of many large corporations, and the major role that this social skepticism has played in the controversy over GE crops, such "democratization" of biotechnology might be as important as biological advances in permitting public approval of GE in agriculture" (Strauss 2003).

A prominent example of the difficulties encountered by scientists working in the public domain is the case of "Golden Rice," which was developed largely with support from the not-for-profit Rockefeller Foundation to alleviate vitamin-A deficiency of children in developing countries. Although the work was carried out in the public domain with an entirely humanitarian aim, more than seventy patents or contractual obligations potentially constrained development of Golden Rice (Potrykus 2001). Thanks to organizational assistance from the Rockefeller Foundation, the private companies holding the intellectual property rights came to an agreement that the needed technology could be used for humanitarian purposes. Today Golden Rice is available free of charge, via national and international public research institutions and local rice breeders to the subsistence farmers in developing countries. A "Golden Rice Humanitarian Board" was established to assist with the next steps in technology transfer and to ensure that the technology reaches the poor that need it most. Breeders are now introducing the Vitamin A trait into locally adapted lines, and numerous researchers are in the process of evaluating the potential health, socioeconomic and environmental impacts. Today, the hold-up on getting Golden Rice into children's and mothers' bellies is not ownership constraints but regulatory obstacles erected, partly in response to fears about GE.

The case of Golden Rice illustrated the need for imaginative ways to address intellectual property issues; ways that do not require years of negotiations, expensive lawyers, or overly complex public/private partnerships to move a crop to market. This is the very large task that CAMBIA and other organizations are now undertaking.

Richard's group has recently succeeded in developing a way to work potentially around a key patented technology used in genetic engineering: *Agrobacterium*-mediated transformation. He and his colleagues found that other species of benign bacteria can be modified in surprisingly simple ways to do the same job (Broothaerts 2005). The resulting gene transfer technology was made available to public sector researchers on an "open source" basis (Dennis 2004).

Richard continues, "The basis of this initiative (called BIOS) is the kernel of the world's first open source biotech toolkit, analogous to the community development that blossomed around the Linux operating system. In the biotech case, the kernel consists of new technologies, such as a new method for transferring genes to plants. Like the Linux open source technology (that is inclusionary, not exclusionary), the BIOS licensee will not receive royalties. Instead, the licensee must agree to share improvements with all other licensees."

Richard is an extremely successful marketer. He has now executed agreements with several nonprofit and for-profit organizations. Recently, CAMBIA and IRRI (The International Rice Research Institute) announced a major joint venture to advance the BIOS Initiative in the hope of using the strategy to galvanize agricultural research focused on poverty alleviation and hunger reduction. "New technologies are increasingly tangled in complex webs of patent and other legal rights, and are usually tailored for wealthy countries and well-heeled scientists," said IRRI's Director General, Robert Zeigler. "Half the world depends on rice as a staple food—but this also means that half the world's potential innovators could be brought to bear on the challenges of rice production, given the right toolkits—and the rights to use them" (CAMBIA and IRRI 2005).

Although the terms of the BIOS license are too restrictive and expensive for many public research institutions (CAMBIA charges and initial fee and asks for "reach-through" rights that allows CAMBIA to share in ownership of any resulting plant varieties) (PIPRA 2006), this kind of creative thinking is needed to move applications of genetic engineering forward for the public good.

Near the end of his talk, Richard asks the audience, "How many got into plant science for money?" There is loud laughter and a few guffaws. He continues, "The answer is nobody. You wanted to achieve benefits for the world and publish widely. The problem is that sometimes your discoveries are patented and exclusively licensed to for-profit companies, and then you can't get them back to modify them or make them available to those that need them. You can't measure success in plant science in terms of accumulation of money. You can measure it in terms of the creation of wealth for society, for example, by enhancing human health. GE has enormous potential for assisting poor farmers in developing countries. Never has the need for an agricultural technology been greater."

Approximately ten years ago, I received first-hand education on the issues of patenting and licensing. It was 1995 and my laboratory team had just isolated the rice gene *Xa21* (Song et al. 1995). At the time, scientists were increasingly patenting valuable genetic discoveries and working with private companies to develop inventions into commercial products.

I had become interested in this gene five years earlier because, on a visit to the International Rice Research Institute in the Philippines, I met Gurdev Khush, a gracious and kind rice geneticist renowned for his contributions to rice breeding. He is a 1996 World Food Prize laureate and 2006 Japan Food Prize recipient. On that visit he told me that *Xa21* confers resistance to one of the most serious bacterial diseases of rice in Africa and Asia, the one caused by *Xanthomonas oryzae* pv. *oryzae*. In bad years, this single disease can cause up to 50% reduction in yield, a huge loss for any farmer. In 1977, his colleague S. Devadath of the Central Rice Research Institute in Cuttack, India, traveled to Mali to evaluate the resistance of an African perennial wild rice species called *Oryza longistaminata* that is a weedy companion of cultivated rice in many areas (Richards 1996). Devadath identified an individual of the wild species that was highly resistant to all tested specimens of the bacterial blight pathogen. He brought the wild rice to the IRRI for breeding studies in 1978 (Khush et al. 1991). Gurdev and coworkers at IRRI introduced the resistance into cultivated varieties using traditional plant breeding techniques (Khush et al. 1991). They found that the resistance was due to a single region on one of the rice chromosomes and named it *Xa21*. Using material obtained from IRRI, in 1990, at Cornell University in the laboratory of Professor Steve Tanksley, I pinpointed the location of *Xa21* on rice chromosome 11 (Ronald et al. 1992). From 1992 to 1995, members of my lab at UC Davis carried out in-depth molecular and transgenic work, and eventually showed that a single gene was responsible for *Xa21* resistance. The work was supported entirely with funds from nonprofit institutions: UC Davis, the USDA, NIH, and The Rockefeller Foundation.

Once isolated, there was tremendous international and commercial interest in using this gene to develop modern crop varieties (Blakeslee, *New York Times*, December 15, 1995; Rundle, *Wall Street Journal*, December 15, 1995). This is because genes are much easier to move into new varieties (either by traditional breeding using modern genetic markers or by genetic engineering; see box 1.2 and figure 4.4) once they have been identified. In addition to improving crop production in rice, many scientists thought that *Xa21* would also be useful for developing new means of disease control in other crops such as the commercially important wheat, maize, and barley. Ultimately, deployment of such engineered varieties could reduce the application of pesticides to the environment. I was therefore confronted with the question of how to further develop this technology for use in crop improvement programs and still make it freely available to less developed countries.

Patents are designed to reward those who make inventive and useful contributions to society, and according to a recent *New York Times* editorial, "Americans think of the granting of patents as a benevolent process that lets inventors enjoy the fruits of their hard work and innovations." The advent of genetic engineering, however, has raised new questions with each passing year. "Should genes be patentable? What about life forms?" (March 22, 2006). In a landmark decision allowing patenting of a living organism for the first time, the U.S. Supreme Court ruled in 1980 (*Diamond v. Chakrabarty* [1980]) that a genetically engineered strain of bacteria that could break down crude oil was a proper subject matter for patent protection under the patent statute. The same year, to promote technology transfer and product development in the United States, the Bayh-Dole Act gave universities and other publicly funded research institutions the right to obtain patents on, and commercialize, inventions made under government research grants (Bayh-Dole 1980).

Patenting continues to play an important role in shaping biotechnology as gene cloning becomes more routine. Whether the principle of patenting genes is morally or ethically correct is a matter of intense debate (Gladwell 1995). There are those who see all biological material as a public good or a gift from nature and, therefore, something that cannot be owned by an individual or company (Gladwell 1995). Others are concerned that patenting will restrict inventions and progress in breeding if germplasm and genes are removed from the public domain. Still, many people see patents as a spur to the process of discovery and development of socially beneficial products and believe that the real ethical lapse would be "for geneticists, having conceived of technologies with vast and immediate therapeutic value, *not* to try to bring them to market as quickly as possible" (Gladwell 1995).

Ingo Potrykus, one of the inventors of Golden Rice, sees it this way: "At one time I was much tempted to join those who fight patenting. Upon further reflection, however, I realized that the development of Golden Rice was only possible because of the existence of patents. Much of the technology that I had been using was publicly available only because the inventors, by patenting, could protect their rights. Without patents, much of this technology would have remained secret or not developed at all without incentives. To take full advantage of available knowledge to benefit the poor, it does not make sense to fight against patenting. It makes far more sense to fight for a sensible use of Intellectual Property Rights" (Potrykus 2001).

UC Davis filed a patent application covering the *Xa21* sequence in 1995 convinced that, without a patent application on file there would not be commercial interest and therefore less overall investment in developing the gene for use in other crops. Although we decided to go ahead with the patent application, it quickly became clear to me that the next step, licensing the invention, needed to be handled carefully. An exclusive licensing agreement with the private sector would eliminate our ability to share this technology with other public-sector institutions, such as national and international research centers that are working on new crop varieties for poor farmers in developing countries. Because rice is the most important staple food in the developing world, improvements in rice yield have a significant impact on global food production. It is estimated that 50% of the potential yield of the world rice crop is lost to diseases caused by bacteria, fungi, and viruses. If the *Xa21* invention was tied up exclusively by one company, the public benefit would be threatened.

We therefore came up with a creative method for licensing. UC Davis negotiated option agreements to license the *Xa21* gene to two companies, Monsanto and Pioneer. As part of the negotiation, the companies were excluded from developing rice varieties in less developed countries. That way, noncommercial researchers, such as those in government-funded programs, would continue to enjoy free access to the genes, so long as they distributed the resulting seeds freely. UC Davis and IRRI formalized this arrangement in an agreement giving IRRI full rights to develop new rice varieties using the cloned *Xa21* gene and freely distribute these new, improved varieties, as well as the cloned gene to developing countries. National breeding programs could then introduce the gene into locally adapted varieties, and be free to distribute these new varieties to their farmers. Because the gene is passed on to the progeny, farmers can grow their own seed for the next season. The new varieties will be genetically identical to the locally adapted variety except for the addition of a single rice-derived gene conferring resistance to bacterial blight. Other traits important for local adaptation (such as drought resistance, hybrid vigor, cold tolerance, or short stature) are expected to remain unchanged. The *Xa21* patent does not preclude the use of *Xa21* by conventional breeding.

Once the exclusivity issue was resolved, I wanted to tackle another, potentially more difficult, issue. As far as I could tell, there seemed to be a paucity of methods to compensate developing nations for their contributions of plant varieties that are critical to the improvement of crops and also critical for the development of new drugs such as anti-cancer medication and antibiotics. The value derived from biological diversity far exceeds the world investment in conservation (Brush 1996).

In cases where plant genetic diversity has been consciously conserved, the rewards have been great. An international system of gene banks has been established that conserves extensively collected germplasm for evaluation and use in breeding programs. For example, the IRRI Rice Germplasm Center preserves 83,000 of the estimated 120,000 rice varieties (IRRI 1990). The benefits to the world community from work at international centers have been "enormous, with low-income food consumers in developing countries receiving the vast majority of those benefits. The total value of germplasm flowing through international research centers to industrialized countries benefited industrialized countries by more than $3.5 billion annually, while the benefits to developing countries for wheat and rice only were approximately $67 billion annually" (Jacoby and Weiss 1997). While conservation and use of plant biodiversity has clearly benefited food production worldwide, a particular country where a specific crop genetic material originated may not have benefited directly.

There is growing concern that industrialized nations, which have the technology and resources to patent and develop commercial products, profit from biodiversity without compensating the providers of the source germplasm. One of the difficulties in assessing appropriate compensation is in predicting that a particular gene will lead to a marketable product. In fact, a particular genetic contribution usually represents only "a small percentage of the total value of the eventual product" (Jacoby and Weiss 1997). In addition, the research and development process required to commercialize a particular product requires enormous regulatory costs, technical knowledge, capital investment, financial resources, marketing efforts, distribution capacity, and time, and is often beyond the budget of developing countries and Western universities.

Because there was no university precedent for germplasm compensation to source countries, and because there was no prior agreement governing intellectual property rights, it was not obvious what would be the most appropriate method to recognize and potentially compensate Mali for its germplasm. What was clear to me, however, was that some form of recognition was needed.

I tried to work through the UC technology transfer offices to develop a mechanism to compensate Mali for its germplasm, but the staff members I spoke to were unsure of how to best make this happen. I was a relatively new professor, unfamiliar with the workings of a large public university, and was stumped. I was unwilling to move forward with the licensing plan until this was resolved. A few weeks later I flew to the Philippines to attend a meeting, again sponsored by The Rockefeller Foundation, to talk about this work. The idea of establishing a compensation mechanism was never far from my mind. As I waited in line to fly home, I saw another attendee just ahead of me in line, John Barton, a courtly, intelligent, and thoughtful Professor of Law from Stanford University. He was also flying to San Francisco on the same flight and we began to talk. It turned out that he was not only an expert on international genetic resources law and technology transfer, but also was quite interested in my dilemma. We talked nonstop during the entire flight home and by the time we got off the plane, it was clear that not only was John willing to help me but also that he was confident we could overcome the impasse. We decided that establishment of a fund dedicated to advanced study or conservation of genetic resources in the developing country would be an appropriate form of compensation. It was likely to be more beneficial to the source nations than a direct financial transfer, because it is usually not possible to determine who in particular should receive compensation as the owner of a specific genetic resource.

With John's help, in June 1996, the University of California at Davis established the Genetics Resource Recognition Fund (GRRF) to recognize contributions of developing nations to the success of UC Davis discoveries (Kate and Collis 1997). The GRRF was to be funded from royalty income generated from commercialization of genetic materials derived from germplasm obtained from developing nations. The goals were to use GRRF funds for fellowship assistance to researchers from developing countries, for farm training projects in the home country, or for conservation of land rich in genetic diversity. The fund was designed to benefit the individuals and farming communities from the same area where the genetic resources were obtained. Students from germplasm-source countries (in this case, Mali) would have first priority. We hoped that the establishment of this program would set a precedent for universities to recognize and compensate for germplasm contributions from developing nations. We also thought that the GRRF would provide a means for scientists to patent their inventions while maintaining productive collaborations and good relations with scientists from developing countries and would create economic incentives for continued sharing of germplasm and conservation efforts.

Although the GRRF makes no effort to assess the future potential income generated from the invention, it provides a constructive solution that would be easy to implement and could be widely accepted. Because it is virtually impossible to predict the commercial success of a single particular invention, ideally the GRRF would be funded from many inventions. Indeed, as of today, no commercial product has yet been made from the *Xa21* gene. Therefore there have been no sales, no royalties, nor funds to distribute. Although disappointing, this outcome was not unexpected. The hope is, however, that as additional UC Davis discoveries are made and licensed to industry, some will find commercial success and the fund will grow over time. Ideally, all future agreements between all UC campuses and companies that license UC inventions could specify a contribution to this fund if the material being licensed is derived directly or indirectly from a developing country. By depositing all the royalties in one fund, the risk that one license may not be profitable would not diminish the overall effectiveness of the fund. Thus each country that contributes genetic resources will benefit from this fund independent of the commercial success of its particular contribution.

Today, the *Xa21* gene has been distributed to 21 countries, as well as to many researchers throughout the United States, and has been widely used in breeding programs around the world. Because we were careful to make *Xa21* available to less developed countries, China was able to move forward in developing hybrid varieties carrying *Xa21*. The addition of *Xa21* to the hybrid parents using genetic engineering did not destroy the value of the hybrid (because only one gene was added). In contrast, traditional breeding would have introduced many genes at once, requiring years to disentangle the genetics before the new hybrid would have been useful. Jia Shirong, a professor from the Chinese Academy of Agricultural Sciences in Beijing said that—after eight years of laboratory trial and field tests—his team had applied to the government for commercial production of Xa21 rice in the central province of Anhui, an area the size of Italy. "The field performance has been excellent," Jia told Reuters in a telephone interview. "Farmers can reduce yield losses and chemical use. Our research data showed that the transgenic rice is as safe as the traditional rice" (Nakanishi 2005). The BIOSafety committee of the Chinese Ministry recommended Xa21 rice for commercialization late in 2004, but it has not yet been released, possibly because of trade problems China could face in light of European consumer opposition to GE plants (Nakanishi 2005).

Our strategy of nonexclusive licensing combined with a contribution to the Genetics Resource Recognition Fund (GRRF) was an appropriate approach for *Xa21*, benefitting both the public and private domains. It does not, however, makes sense for all genes. For example, after some consideration, UC Davis did not file a patent application on the *Sub1* gene (see chapter 1) because the immediate need for this gene was primarily for rice in the developing world. Furthermore, the likelihood of generating a commercial product in other crops would require years of additional and expensive research so it seemed unlikely that a profit would be made that would benefit UC Davis or the people of Orissa through the GRRF. We therefore concluded that the public would benefit most broadly if we rapidly placed the *Sub1* gene in the public domain.

Gary Toenniessen, a deep-voiced, understated man with a great sense of humor, and a seemingly endless capacity for travel, piloted the Rockefeller Rice Biotechnology Program from its inception in 1985 to its completion in 2000. Trained as a microbiologist, he was responsible for developing and implementing programs that would help address environmental problems associated with farming. This included work on alternatives to persistent pesticides and improved management of agricultural resources.

The community of rice researchers created and supported by Gary and the Rockefeller Foundation was phenomenally successful in carrying out the mission stated early in the beginning of the program, and it left a striking legacy: the complete sequence of the rice genome, development of simple transformation technologies for rice that can be used throughout the world, and creation of more prolific, robust, resistant, and nutritious strains.

Most, if not all, rice researchers want to ensure that GE breakthroughs and useful technologies benefit less developed countries and small farmers in rice and other crops. Several years ago Gary and his colleagues at the Rockefeller Foundation joined together with the McKnight Foundation, both of which support plant biotechnology research in developing countries and with several of the leading agricultural universities and plant research institutes in the United States including the University of California, Cornell, Michigan State, University of Wisconsin, North Carolina State, University of Florida, Ohio State, Rutgers University, Donald Danforth Plant Science Center, and the Boyce Thompson Institute. Their goal was to establish a Public Intellectual Property Resource for Agriculture (PIPRA) to reinvigorate the linkages between universities and the international agricultural research system, and to build new partnerships with the private sector to take advantage of their expertise and resources (Atkinson et al. 2003).

PIPRA allows universities to market their technologies to the private sector, and thus still profit from their inventions, while retaining rights for humanitarian purposes and small crops that are vital to small-acreage farmers; the goals are similar to those of CAMBIA but with fewer restrictions (http://www.pipra.org). Along with Toenniessen, Alan Bennet of UC Davis was a major driver of the development of PIPRA, and PIPRA is now located at UC Davis.

It is now March, typically a cool and sunny time in the valley, but it has been raining for the past month. Many counties in Northern California have been declared to be in a state of emergency because of flooding. As I look out my window wondering if I should skip my daily trip to the pool and swim home instead, I hear a knock at the door. Rebecca, my hardworking and creative assistant, walks in. She takes a look out the window and says, "Isn't California supposed to be a sunny state? I mean when my roommate convinced me to move here she told me it would be sunnier and warmer than Wisconsin. It's more like being back in Burkina Faso in the middle of the rainy season."

I knew that Rebecca served in the Peace Corps as a health volunteer in sub-Saharan Africa before moving out to Davis, but we had not yet had a chance to talk about her experiences. I thought her comment was kind of funny because Burkina Faso was known to be hot and dry. So I ask for an explanation.

"After the scorching, stagnant hot season towards the end of May, if we were lucky, it would rain like this until October," she said, gesturing to the window. "I used to wish for rain almost everyday, but it is not the first thing that comes to mind when I think about Burkina."

"What is?" I ask.

"The heat and poverty. When I first I stepped off the plane and onto the tarmac the heat still hit me like a wall. That evening will be imprinted in my mind forever—after a brief greeting with other volunteers and staff, Peace Corps ushered us new volunteers into a few cars and drove us through Ouagadougou, Burkina's capital. I couldn't take my eyes off of the streets; there were hundreds of stands set-up along the road, mostly made out of wood and a tin roof, with people selling everything from meat cooked on a stick to mangos and other fruits I only rarely saw in my local supermarket. And then there were the kids—in tattered clothes that looked to have come from second-hand shops in The States—running through the streets barefoot."

She tells me that there are over 14 million people living in the country that's only the size of Colorado—a state with about 4 million people. "What shocked me the most was the raw poverty, with kids selling Kleenex or chewing gum for ten francs, the equivalent of two cents. And the smell of the city—an odor like pungent French cheese mingled with sweat, cooking meat, spices, and a bit of sweetness from the fruit stands. People ate a lot of tô, a gelatinous substance made from millet flour. Although some people in my village had side jobs, like selling things in the market and such, the families was essentially subsistence farming. They mostly grew millet, sorghum, maize, peanuts, and beans. Women would take the millet stalks and pound them in a huge mortar and pestle, so the seeds could be easily ground into flour with flat rocks. They would stir the millet flour into boiling water, and then whip it into a paste; it hardens as it cooks so that when it cools you can pick it up in your hand and mold it into a ball. It was this constant pounding of millet that I called the heartbeat of Ridimbo, my village."

She adds that getting enough protein was a problem, especially for children. Families would often add some dried fish that they bought in the market to their nightly tô, sauces, or perhaps some chicken if there was a guest. Goat was available a number of special occasions, but sheep and beef were scarce, as they are more expensive and much harder to come by. If meat was available, a guest would get the first choice, followed by the man of the house, then the women, and whatever was left would be for the children. For the most part kids got enough calories, tô is good for that, but they didn't get enough protein or vitamins. Consequently she saw many kids with swollen bellies and reddish hair, the telltale signs of the disease kwashiorkor, malnutrition that is due to a lack of protein.

"Do you think that the people of Ridimbo would be interested in a GE protein-enriched millet or vitamin A-enriched rice?" I ask.

"They would appreciate learning about these GE foods, and how they would affect their health. I spent a number of days each year biking from village-to-village in my health center's coverage zone, in order to vaccinate kids for polio and hand out synthetic Vitamin A supplements, as apart of the national campaign. Over and over again on these trips I heard the same phrases "La santé avant tout," or *health first* and "Laafi beeme," which in Mooré, the language of the Mossi people, means *health is here*. In villages where cultivating your field is painstakingly slow with a small, hand-held hoe, being in good health is not taken for granted. In fact, "Laafi" or *health* is the response to all daily greetings. Also, I think that they would be interested in learning about high-yielding crops, as it could reduce their workload tremendously. But if there is one thing I learned while I was working there, it is that no matter how wonderful the technology or ideas you would like to implement could be, how good your intentions are or how great you perceive the need to be, ultimately it is the people, the consumers, that need to make the decision. You can distribute free seed, which works well in experimentation trials, but if they do not yield well in the farmers' hands or if there is different maintenance involved that people are not expecting, then they may not be willing to continue using the products regardless of the theoretical benefits."

This comment reminds me of a classic case that Leland Glenna a professor of sociology at Penn State, who earned a master's degree at the Harvard School of Divinity before earning a Ph.D. in the sociology of agriculture and natural resources, once described to me. "A number of years ago in East Africa, farmers tried a dwarf wheat variety that increased yield. After a year or two, the farmers went back to their old variety. Researchers were baffled until a social scientist went out and asked the farmers why. It turned out that the dwarf variety may have improved yield in experiment station trials, but it did not in farmer fields. The dwarf wheat stems were stronger, so birds would perch on the wheat stems and eat the grain." Clearly genetic modification alone (whether by traditional breeding or GE) is not necessarily a silver bullet for solving complex hunger problems. It has to work in the hands of the users, the poor farmers.

This conversation with Rebecca reminds me of the situation of poorest of the poor. How can we make GE technology available to those who most need it, and how can we be sure that the technology is directed to that which will be useful to them?

The Rockefeller Foundation, under the direction of Gary, is now working to address just this problem. He is currently leading the Foundation's work aimed at improving food security in Africa, and has helped establish the African Agricultural Technology Foundation (AATF). Because local organizations are the most able to determine and develop what is relevant to the needs of their consumers, the AATF identifies African organizations that would like to utilize publicly available materials, and links them with private institutions that could further help them develop new crop varieties, conduct appropriate Biosafety testing (which is currently exorbitantly expensive), distribute seed to resource-poor farmers, and help create local markets for excess production. To date, a number of international seed companies and the U.S. Department of Agriculture have expressed interest in working with AATF (http://www.aftechfound.org).

Genetic engineering is an appropriate tool for many situations, but not all. In sub-Saharan Africa for example, where depletion of soil fertility is a major biophysical cause of low per capita food production, it will be necessary to first restore the health of the soil before the introduction of new varieties, GE or non-GE, will have much of an effect unless there is one day a GE crop that makes better use of soil nutrients or fixes nitrogen (Sanchez 2002; Oldroyd 2006). A report from the International Center for Soil Fertility and Agricultural Development (IFDC) in the United States says that Africa's soils are being stripped of nutrients so fast that 75% of farmland is now severely degraded. "When farmers plant the same fields season after season and cannot afford to replace nutrients taken up by their crops, the soil is literally mined of life," says IFDC President Amit Roy (Henao and Baanante 2006). Fertilizer use is lower in Africa than anywhere else, amounting to less than ten percent of the global average. This is because fertilizers cost two to six times more in Africa than elsewhere, a price that is too high for most farmers there. On-farm organic production practices that involve using nutrient-building plants or applying just minimal amounts of fertilizer precisely where they are needed can be an effective, low-cost strategy to increase yields (Barhona and Cromwell 2005). Only after the soil's fertility is replenished can the much needed, high-yielding crops thrive.

Without good government, establishment of policies directed at promoting food production (rather than just commodity production), appropriate intellectual property policies that assure agricultural technology is publicly available, and fair international trade policies, a discovery that may help farmers and consumers around the world is useless. GE may help Burkina Faso, but only within the framework of a viable public sector that promotes the public good.

Leland agrees with this concept. "The issue is to create an agricultural and food system that is directed broadly at the public good, not one dominated by private interests. This may lie at the heart of the distrust between policy makers, farmers, consumers, university plant breeders, and industry plant breeders. GE is not the problem, per se, from my perspective. The problem is that concerns about potential overreaching by private interests have completely overwhelmed the discussions on how the public good can be achieved. And GE seems located in the center of the tradeoffs because of the intellectual property issues that are so intimately connected with GE. The question about the public good is being lost in the debates over pollen transfer, intellectual property, farming systems, and food safety. These are broader, political-social issues. They are important, but that will arise even in the absence of GE" (L. Glenna, personal communication).

Through Gary's work and projects such as PIPRA, the GRRF, AATF, and BIOS, creative approaches are increasingly being developed to achieve this public good, both within and outside of the United States. Our farms need these tools. "The alternative is deterioration in the world's ability to cope with the problems of hunger and disease" (Jacoby and Weiss 1997).

At the end of Richard's speech in Seattle to the group of plant biologists, he exhorts the audience to become activists, to ensure that their discoveries are made widely available so that they can be used to improve agriculture, as well as the lives of those less fortunate. "If plant geneticists had pushed this open source approach years ago, I would still be able to get a free beer."

Dinner

Twelve

Deconstructing Dinner: Genetically Engineered, Organically Grown

Instead of indulging into a fruitless debate about what strategy would be appropriate in agriculture, it would be much more rewarding in looking at the best way forward for a given country, a given ecology and economy. Looking for sustainable and equitable farming methods means . . . to refrain from any kind of ideological debate and concentrate on pragmatic decisions to find the best solution for a given region. Roads to success . . . are many, and we must pursue them all.

> Klaus Ammann, Director, Botanical Garden, University of Bern;
> from "Sustainable Food Security for All by 2020," Sept. 2001

The study of connections is an endless fascination, and the understanding of connections seems to me an indispensable part of humanity's self defense.

> Wendell Berry, farmer and former Professor of English at the
> University of Kentucky; from *Home Economics: Fourteen Essays,* 1987

School is out with a clang of a bell and an explosion of motion. Homework and lunch boxes are thrust into the waiting parents' hands and quickly forgotten; shoes are kicked off. It is a warm spring day and time to move beyond learning to a more essential activity—eating. Cliff pulls on my hand, interrupting the good-byes, and we walk home. Our house sits across the street from the two-room school, in a dusty part of town. The house, with faded cedar shakes and maroon trim, resembles a craftsman-style bungalow. In back, adjacent to the tiny organic farm, is a large barn that houses Raoul's boat-building shop. The enormous outdoor wall is painted with a colorful whimsy of plants and animals with a DNA helix in the background.

Cliff skips up the front path bordered on both sides by a garden that is soaked in a fragrant color and scent, bred to delight the senses: pink blooms of evening primrose, yellow hooded petals of Jerusalem sage, and small purple flowers of fragrant rosemary. He stops abruptly, noticing that the fresh green leaves of the evening primrose are tramped down, and swerves off the path to investigate.

"Mama! I found a nest, and it has eggs in it! What kind of eggs, Mama?" Cliff yells excitedly, hoping for something exotic.

I look closely. The eggs are suspiciously large and bluish suggesting that one of our Ameraucana hens, a breed developed in the 1970s to incorporate the favored "blue" genes from a South American bird, has escaped again, maybe planning to start a family. But she is not off to a good start—the pastel-colored eggs are cold and abandoned.

"I think one of our hens is loose," I answer.

Cliff, with a slightly disappointed look, picks up the eggs and asks, "I'm hungry. Can I eat them?"

"Sure." We go into the house, where I fry them in butter with salt. I toast a slice of homemade wheat bread made with walnuts grown on our friend Paul's farm (recipe 12.1). I slather the toast with butter and Grandmother Ronald's homemade plum jam (recipe 12.2). A perfect after-school treat for a hungry boy.

RECIPE 12.1

Pam & Trish's Whole Wheat Bread with Walnuts

(2 Loaves)

INGREDIENTS

1 c Cracked wheat cereal or Jennifer's cereal mix (a blend of grains: hard wheats,
   soft wheats, barley, and rye—see Jennifer's description below)
3 1/4 c Water
1/3 c Honey
3 Tbsp Butter or oil (we prefer half oil and half butter.)
2 tsp Salt
3 c Whole wheat flour
2 Packages of yeast (4 1/2 tsp)
1/3 c Instant nonfat milk powder
3-3 1/2 c All-purpose flour
1 c Chopped walnuts

1. In medium saucepan combine cracked wheat and water, bring to a boil.
2. Reduce heat, cover and boil for 8 minutes. Add salt, honey, and butter; cool to lukewarm.
3. In mixer bowl combine whole wheat flour, yeast, and milk powder; add cooled cereal mix; and beat at low speed for 30 seconds.
4. Beat at high speed for 3 minutes. By hand stir in enough all-purpose flour to make moderately stiff dough. Turn onto floured surface.

5. Knead until smooth and elastic for 8–10 minutes. Add walnuts and shape into a ball. Place in greased bowl, turn once to cover surface.
6. Cover and let rise in a warm place until double in size, 45–60 minutes.
7. Preheat oven to 400°F.
8. Punch down risen dough, divide in half and cover. Let rest for 10 minutes.
9. Shape into loaves, place into greased pans. Cover and let rise until double in size, about 30 minutes.
10. Bake at 400°F oven for 30–35 minutes. If loaves brown too quickly, cover with foil.
11. Remove from pan and let cool on racks.

To serve, toast slices of bread and spread with butter and honey or plum jam.

Jennifer's cereal mix: "this year I grew out a huge variety of heirlooms in one planting all mixed up that I'm gonna use in a cereal mix. So I can save the seed without separating it all out. Sort of like gene cesspool cereal mix."

RECIPE 12.2

Trish's Plum Walnut Jam

INGREDIENTS

4 c Sliced Santa Rosa plums
3 c Sugar
1 c Chopped walnuts (from Terra Firma Farm)
1 tsp butter

1. Bring plums and sugar to a full boil. Boil until jam clings to spoon and a drip gels when it hits ice water and holds it shape.
2. Add chopped walnuts and bring back to boil for 1 minute then turn off.
3. Skim foam off top and add 1 tsp butter to jam.
4. Immediately put into clean jars. Screw on tops and let cool.

Jam jars should be stored in freezer.

In California's Central Valley, food cannot always be scooped up from the front path on the way home from school, but nearly. Here food is abundant, and it is fairly easy to figure out what to eat, especially if you are not overly concerned about the presence of GE ingredients or the wanderings of a genetically modified hen that lays blue eggs. If there is meaning to be found in each meal, it is not about how the food was genetically modified, but in the freshness of ingredients, the health of the farm workers, the impact on the environment, and the mood and gratitude of the diners.

Even the most basic foods must be cultivated carefully and thoughtfully, and the problems confronted by farmers need to be creatively addressed. Take the walnuts in our bread, for example. A few months ago I toured a 320-acre local walnut farm. Michael Pollan, the author of *Omnivore's Dilemma*, a journalistic investigation of how food gets from the farm to the table in the United States, was visiting that day with his students from UC Berkeley. The Yolo county farm advisor, Rachel Long, was there too and showed us how to set pheromone traps to attract codling moths that were attacking the walnut trees. The idea is to confuse the males so they cannot mate with the females. As we were learning to place the traps up high, Michael asked Craig McNamara, the farm owner, the age of his walnut grove. For walnut trees they were quite small—and were seemingly just a few years old.

"Funny you should ask," responded Craig. "These trees are already fifteen years old, they are stunted and weak, not because the moths are so much of a problem anymore, thanks to Rachel's biological control using these synthetic pheromones." He goes on to explain that the problem is that the soil is heavily infested with nematodes, which are microscopic worms. "There are no organic methods to control these nematodes, and because we did not want to fumigate the soil with methyl bromide before planting, we accept the reduced yield."

There is good reason not to use methyl bromide. It kills every living thing in the soil, renders farm workers at high risk for prostate cancer (Alavanja et al. 2003), and is a chemical that contributes to the depletion of the Earth's ozone layer. Despite these alarming facts, the United States applies tens of millions of pounds of methyl bromide each year—most of it used to fumigate soil before planting crops, which is why we sometimes see, early in the season throughout the valley, tarps covering acres of earth to prevent escape of this gas (Pesticide Action Network North America 2005).

The nematode problem is a good example of how genetic engineering could potentially benefit farmers, consumers, and the environment. Virtually all commercial walnut trees are genetic chimeras; that is, the lower trunk and roots are from one species (California black walnut), that is resistant to a serious fungal disease, whereas the scion (nut producing top part of the tree) is from another species (English walnut) that produces the nuts that consumers prefer. This type of biological technology (mixing of two species in one tree!) is allowed under the USDA National Organic Program.

Researchers at UC Davis are investigating the possibility of genetically engineering the rootstock of the walnut tree with a piece of DNA that would "silence" an essential gene that the nematode needed for its survival (box 12.1). The idea is that the when the nematode sips on the cell it will also suck in this silencing construct. A few hours later, the nematodes will die (V. Williamson, personal communication. Sept. 5, 2006). If this approach works, then an English scion could be grafted to the newly resistant rootstock, producing walnuts with no "foreign" genes. I timidly asked this organic-oriented group what they thought about the idea of using GE to combat this pest.

Rachel said, "That is a very interesting idea. Overall, I am excited about the possibility of using GE as a tool to reduce toxic pesticide runoff into streams. Water quality is an important issue for me." The others are not so enthusiastic. Craig paused and then said, "I don't know what to think about GE."

I took that brief hesitation and acknowledgment of uncertainty as a sign that organic farmers are increasingly interested in evaluating whether or not GE can benefit their own farms. No one in the group seems to think that GE is a bad idea in general, and I am encouraged, because I don't want to see the ecological farming community cede this useful technology to others. The current focus on the *process* of how a new variety is created (manual pollen transfer, grafting, mutagenesis, or genetic engineering) seems to be a distraction from the promotion of activities that would help growers farm more ecologically.

BOX 12.1 **RNA Interference: Gene Silencing**

RNA interference occurs in plants, animals, and humans. In addition to its use in engineering immunity against pests of plants, such as papaya ringspot virus (see box 4.2; Gonsalves, 1998), this approach has also been used to develop a treatment for macular degeneration, a disease of the eye that is gradually robbing Raoul's father of his eyesight (Campochiaro, 2006; Shen et al. 2005).

The fundamental importance of this technique as a method to study the function of genes, as well as for its promise for leading to novel therapies in plants and animals was recognized by the Nobel prize committee in physiology or medicine who awarded the 2006 prize to the American scientists Andrew Fire and Craig Mello for this technique.

Genomes operate by sending instructions for the manufacture of proteins from DNA in the nucleus of the cell to the protein-synthesizing machinery in the cytoplasm. These instructions are conveyed by messenger RNA (mRNA). In 1998, Fire and Mello published their discovery of a mechanism that can degrade mRNA from a specific gene. This mechanism, RNA interference, is activated when RNA molecules occur as double-stranded pairs in the cell. Double-stranded RNA activates biochemical machinery that degrades those mRNA molecules carrying a genetic code identical to that of the double-stranded RNA. When such mRNA molecules disappear, the corresponding gene is silenced, and no protein of the encoded type is made.

At home, a few hours later, the front door opens and Raoul walks in, arms full of a large basket of freshly harvested kale, greenhouse tomatoes, lettuce, green garlic, and herbs.

"What shall we have for dinner?" he asks jokingly, because it is apparent that the answer lies within the basket. We like the variety, freshness, and ease that comes from eating off the farm. At home on weekdays, we like to prepare the food quickly, and we want it to be colorful (we figure if there are a variety of colors on the plates, then we are getting enough vitamins) and tasty.

"How about tofu tortillas?" he asks.

That sounds fine to me, so I open the refrigerator and pull out some tofu. I remember that we need to eat the asparagus in the garden before they mature into inedible, fibrous stalks. "Raoul," I ask, "would you please pick some asparagus for dinner?" He walks out the back door and quickly returns, hands full of fresh green spears. I rinse the asparagus, the lettuce (tossing out a slug), and the tomatoes. Then I spin-dry the lettuce and begin to chop up mounds of kale. I sauté the kale and asparagus with garlic, chile, and salt.

For the tofu tortillas, we prepare a variation on one of Raoul's recipes (recipe 12.3). I grab a knife and slice the end of the plastic bag holding the tofu. As the water runs off the chopping block and onto the floor, the children elbow their way in and say, "Slice me a piece, Mama." I quickly wipe up the mess and slice off a thick piece for each child. They drip their way to the table and start munching. I clatter around the cupboard looking for the grater; it tumbles out of the over-packed cabinet, adding one more ding to the old pinewood floor. I pick it up and grate the block of tofu, shredding it into white confetti. The tofu is made from certified organic soybeans that are processed into the familiar white square blocks in a factory a few miles away.

After I finish preparing the tofu, I turn on the stove to high and pour in a few tablespoons of olive oil. The oil is not certified organic but it was on sale and locally made. The low cost appeals to me and so does eating locally grown foods because it supports our neighboring farmers. The olive oil is definitely not GE because there are no GE olives on the market. Despite this fact, the label on the bottle says "GE-free." It is a hopeful marketing ploy that is often seen at our local food co-op where many consumers associate GE with massive farms, pesticide runoff, and high fertilizer use. Yet genetic engineering is not the cause of these types of farms. The industrialization of agriculture, with the associated high inputs of pesticides and fertilizers, proceeded quite contentedly for years before the advent of GE, fueled mainly by governmental agricultural policies that do not put high priority on social and environmental costs. Ironically, much of the food labeled "GE-free" may have been imported from afar, grown with toxic pesticides, or be less nutritious than the local fare. In contrast, food that is GE may be locally grown without pesticides, and someday, be more nutritious than crops grown from non-GE seed.

RECIPE 12.3

Pam & Raoul's Tofu Tortillas

INGREDIENTS

12 Corn tortillas
2 c Gruyère or cheddar cheese, grated
2 Garlic cloves, smashed and chopped
1 Tbsp Cumin
1/2 tsp Chile flakes
1 lb Firm tofu, grated
1/4 c Sunflower seeds
3 Tbsp Soy sauce

1. Fry tortillas on both sides in olive oil. Sprinkle on grated cheese. Keep warm.
2. At the same time, in another pan, fry garlic with cumin and chile. Add grated tofu, sunflower seeds, and soy sauce. When brown, fill tortillas.
3. Serve with fresh salsa, avocados, greens, and chopped tomatoes.

"Bang!" I wield an enormous knife blade to smash a clove of garlic—a trick I learned from the mother of my first student, Wen-Yuan Song, on a visit to his home in China years ago. I mince the smashed clove and toss it into the hot oil, add a tablespoon of cumin (likely imported from India or the Middle East and reputed to keep chickens and lovers from wandering too far), and dump in the grated tofu. A few minutes later, once the onions and garlic are soft, I add a handful of sunflower seeds and soy sauce. The soy sauce, too, is not organic and is likely made from GE soybeans that contain trace amounts of a bacterial protein that protect the plants from the herbicide glyphosate (see box 5.2). No matter, it is drizzled in anyway. After all, the bacterial protein is not toxic to humans and if there is any left after the soybean processing, it will be quickly denatured in the heat of the pan.

I pull out another pan, turn a burner to high, pour in some more oil and plop down two of Micaela's corn tortillas, made in a factory ten miles north of here. The ingredients are simple: corn, water, and salt. The corn is not certified organic and the tortillas likely contain trace amounts of Bt protein. We choose them because these are the best tasting tortillas in the world and they are produced close by our home.

It seems to me that these tortillas made from corn from Bt-corn plants fit well within the ecological farming framework we try to support. First, the global environment is being spared more than a hundred million pounds of much more toxic pesticides each year (Toenniessen 2006). Second, the tortillas likely contain reduced amounts of mycotoxins as compared to tortillas made from conventional or organic corn (Kershen 2006; see box 5.2).

In the summer, if Pam and I were to prepare this meal, we would almost certainly include some fresh-picked sweet corn with their wormy ends chopped off. Cutting off the tips doesn't bother us and most of my customers tolerate the worms. I have wondered, however, if consumers would prefer a wormless GE ear.

To begin to answer this question, Pam and I surveyed undergraduate students in a science and society class at UC Berkeley last year. We had explained that Bt toxin is a protein that has been used to engineer pest-resistant plants. At the end of our lecture we asked the students if they would prefer to eat: (1) Sweet corn grown with synthetic pesticides and fertilizers (no earworms); (2) Sweet corn grown organically (with earworms); or (3) Sweet corn grown organically containing genetically engineered Bt toxin (no earworms). Out of a class of 25 students, 22 voted for number 3. I was surprised because I thought students from UC Berkeley, known for their activism, would be strongly opposed to GE foods. But this class was full of science majors, and they believed that the science behind GE crops was sound. They also liked the environment-friendly approach of organic farming; the combination of GE and organic farming seemed like the best strategy for the future of agriculture. Interestingly, when Pam gave the same survey to students in Davis it yielded a similar result—organically-grown Bt corn was the top choice. We later discovered a more rigorous study conducted by Doug Powell at the University of Guelph (Powell et al. 2003) showing that consumers preferred GE corn over non-GE corn when clear labels and explanations were used. In other words, many consumers will choose GE if there are clear aesthetic and environmental benefits.

"Ahhh." It is starting to smell good and the kids are hungry, so Pam leaves the stove, slices some more tofu, fills a couple glasses with milk and walks over to the children. While she is occupied, the oil begins to smoke and the tortillas are becoming overly-crisp. I hurry over to turn down the heat on both burners, and glance at Pam wondering why she has not yet learned from the hundreds of other meals she has burned in this way.

"Whoops, sorry!" I say as I rush over to flip the tortillas. I know Raoul does not like my multi-tasking approach but usually, if I time it just right, I can do something else and also get the tortillas flipped before they smoke. I quickly finish grating piles of Gruyère cheese and drop a handful on the tortillas.

Raoul prepares a dressing (olive oil, red raspberry vinegar, salt, and pepper) and rips up the salad greens. He grates the carrots and breaks up the goat cheese on top.

We belong to a subscription grain delivery service that brings us flours, beans, and goat cheese every month. The farmer, Jennifer, used to work with Raoul at Full

Belly Farm and now has her own farm three hours north of here in a spectacular valley nestled beneath the foothills of the Trinity Alps. She gives loyal customers the opportunity to buy a goat or two in trade for cheese. On our first visit to the farm we purchased Lucy, and today we are eating her cheese. It's delicious.

As I work on the salad, Pam pulls some hybrid Cobra tomatoes out of the basket—our first this season, grown in the greenhouse at the student farm so that my customers will have an early spring treat to tide them over until the field tomatoes are ready in July. In Davis, tomatoes are usually not planted until after the last frost, which can be as late as April 15th, because tomato plants die when the temperature drops below freezing. Although I already knew this, I relearned this lesson the hard way the first time I transplanted tomatoes into the field behind our house. It was early April and I had calculated that the risk of a frost was quite low and the probability of bringing in good money on an early tomato crop was quite high. So much for farmers' calculations. On April 10th, the weather station announced that a cold front was moving in so I covered the seedlings with floating row cover, hoping that a bit of extra protection would keep the tomatoes a couple of degrees warmer and that this would be sufficient to keep the seedlings alive. Lacking a way to sprinkle water on the plants, which also helps protect from frost, there was nothing to do but go to bed and hope for the best. First thing the next morning we ran outside to look. The ground looked like a cake covered with powdered sugar—frost everywhere, the whiteness contrasting with the black withered leaves of our tomatoes. A few weeks later I assessed the damage. We had lost about half the crop and those plants that survived grew back slowly. I was not the best predictor of weather that year, but fortunately for us it was not a life-or-death situation.

As Raoul knows, this of course is not the case for the vast majority of farmers on the Earth, where tolerance to environmental fluctuations such as cold, salt, or submergence can mean the difference between eating or not. Traits such as these are the most difficult to address using standard breeding approaches. In the future, this is where genetic engineering will likely have the most significant human impact. There are already examples of GE plants that can tolerate very cold temperatures (Thomashow 2001).

I slice the tomatoes, the red juice puddling with the tofu water. I take a piece and pop it in my mouth. Sweet and tangy, the taste evokes the farm and the earth; and the beginning of summer. As soon as there are some frost tolerant varieties that taste good and thrive in our backyard, I will encourage Raoul to buy the seed, GE or not. Eating locally, when possible, generally means the food is fresher and more nutritious. If breeders and geneticists can come up with plant varieties that are tolerant to cold, we can extend the season for eating fresh, locally grown produce, without using a lot of energy growing tomatoes in greenhouses.

Teaching Sang Min about organic farming last summer reminded me that a farmer should try, even though it's difficult hearing it from some people, to entertain new ideas and techniques and to seek out the most appropriate, environmentally safe technology to tackle a particular problem. As Mike Madison, a fellow farmer and writer says, "In dealing with nature, to be authoritarian is almost always a mistake. In the long run, things work out better if the farmer learns to tolerate complexity and ambiguity . . . Having the right tools helps" (Madison 2006).

Is GE simply a new tool for farmers that in some cases will be the right one? Although genetic modification by conventional breeding and genetic engineering methods are distinct processes, they ultimately have the same end—to alter and improve the genetic makeup of a plant. Whether GE plants fit into a framework of ecological farming gets back to the first thing I tell my students: Organic farming is about health—health of the soil, the plants and animals, the farmer, the consumer, and the environment. A marriage of farming with biological science has always been an important strand of the organic approach. Plants that have been genetically modified using older methods have given rise to nearly every food we eat. Such crops, which are resistant to diseases, insects, or nematodes, fit in well with organic production, and it seems to me that there is a role there for the right GE plants as well.

At the same time I think that much of the potential of GE plants is lost in conventional systems that continue to use pesticides and synthetic fertilizers. To maximize the benefit of GE plants they would best be integrated into an organic farming system. In this way there is a complementation of practices and technology—the organic practices protect the environment and the GE technology helps reduce crop losses to disease or environmental stress.

Dinner is ready. Raoul, the children, and I sit down and each take a corn tortilla with melted Gruyère and load it up with the garlic-flavored, cumin-infused tofu, avocadoes, Micaelas' salsa, chopped tomatoes, and fresh greens. We serve the kale-asparagus dish on the side.

I pour the water and wine. The children pick up their stuffed, dripping concoctions.

"Wait! Who is going to say the prayer?"

"I will," says Audrey. They wipe their hands and hold each others and ours.

"Thank you for this lovely dinner, and I hope everyone in the whole wide world gets better soon."

The children stand up on the chairs, yell "hip hip hooray" and then settle down to eat.

In what seems like a second later, there is a call for dessert. Tonight's dessert is plum cake, a recipe of my aunt's (recipe 12.4).

RECIPE 12.4

Tante Lissy's Pflaumenkuchen (Plum Cake)

INGREDIENTS

1 c Butter
1 c Sugar
1 Egg
2 tsp Almond extract (or vanilla)
1 tsp Salt
1 c White flour
1 c Barley flour
10 Plums, pitted and cut in half
2 Tbsp warmed apricot jam

1. Beat together butter and sugar. Add in egg, almond or vanilla extract, and salt.
2. Mix in flours to form a dough.
3. Pat 2/3 of the dough into an 8-inch pan with removable rim. Arrange plums, cut side down, in pan.
4. Lattice rest of dough on top; drizzle with apricot jam.
5. Bake at 350°F for 45 minutes.

I saved some Santa Rosa plums last summer and froze them for just such an occasion. We are lucky to have an orchard with plenty of "stone" fruits such as apricots and peaches, and I hope that they will always thrive here, but I am not sure. Stone fruits are susceptible to plum pox virus (PPV), which has been a devastating disease in Europe since the early 1900s. In 1992, PPV was reported for the first time in Chile, and in 1998 was found in an Adams County, Pennsylvania orchard. Although the disease remains localized at this time, the only known method of control, in the case of an outbreak, is to pull up the trees and bulldoze them before the disease spreads to other parts of the Americas (The Plant Disease Diagnostic Clinic 2001). Because of this threat, the U.S. Department of Agriculture developed a GE plum variety that is resistant to disease, applying a similar technique that was used to engineer papaya for resistance to papaya ringspot virus (see box 4.2). The GE trees look like their non-GE female parent—Bluebyrd—a commercial cultivar developed through traditional breeding. And their fruit tastes the same. In an interview with ARS staff, horticulturist Ralph Scorza said, "It's basically immune to the plum pox virus. We've shown that it is resistant to all major strains of the virus that we've tested" (McBride 2006). Recently the USDA Animal and Plant Health Inspection Service (APHIS) announced that it has "deregulated" the GE plum variety, which brings it a step closer to commercial cultivation (Iadicicco and Redding 2006; Kaplan 2007).

The U.S.-based Organic Consumers Association has already come out against the new plum, arguing that approval of the GE plum variety would "open the door" to authorization of other GE stone fruits (Organic Consumers Association 2006). And they are probably right. If the GE plum is popular, peach growers will likely also want to use the technology, because although strict preventive measures have kept plum pox out of the state so far, experts say it could still sneak into small, backyard orchards (Haire 2001). Phil Brannen, an Extension Service plant pathologist with the University of Georgia College of Agricultural and Environmental Sciences says that "Plum pox virus can devastate entire peach orchards. Infected trees produce unsweet, blemished fruit that can't be sold. And once a tree is infected, there is no cure."

Advances in genetics have been fueled by intense scientific curiosity about basic aspects of biology. New technological breakthroughs have accelerated discoveries. For example, we now know the genome sequences of some plants, as well as the DNA sequence of many of the microbes that infect them. And we also know that scientists and breeders can use this information to develop biologically–oriented, sophisticated, and elegant approaches to address agricultural problems.

There seems to be a communication gap between organic and conventional farmers, as well as between consumers and scientists. It is time to close that gap. Dialogue is needed if we are to advance along the road to an ecologically balanced, biologically based system of farming. Science and good farming alone will not be sufficient to provide food security to the healthy, or to the poor and malnourished, or to solve all of our current ecological woes. Governmental stability, as well as governmental policies, plays a large role in ensuring food security. Without science and good farming, however, we cannot even begin to dream about maintaining such a secure future. Genetic engineering is not a panacea for poverty, any more than conventional breeding is or organic practices are, yet it is a valuable tool that farmers can use to address real agricultural problems such as pests, diseases, weeds, stresses, and native habitat destruction. Like any tool, GE can be manipulated by a host of social, economic, and political forces to generate positive or negative social results.

Presently, the vast majority of commercial GE crops are either those that carry the pest-killing Bt toxin or those that carry tolerance to Monsanto's herbicide glyphosate (Roundup). Only a handful of others are on the market, yet the potential beneficial applications of GE are vast. It seems nearly inevitable that genetic engineering will play an increasingly important role in agriculture. The question is not whether we should use GE, but more pressingly, *how* we should use it—to what responsible purpose. Consumers have a significant opportunity to influence what kinds of plants are developed. Agriculture needs our collective help and all appropriate tools if we are to feed the growing population in an ecological manner. Let us direct attention to where it matters—the need to support farming methods that are good for the environment and for the consumers. Rather than focusing on the unforeseen consequences of combining *Prunus domesticus* (originally from Japan) with snippets of an edible virus, I prefer to "celebrate the triumph of human ingenuity" (Moses 2006) that will allow me to continue to bring Aunt Lissy's plum cake to our table.

Our children, anyway, are not thinking about these details. They are finishing their dessert. After all, this is about eating and eating well.

Glossary

AATF   African Agricultural Technology Foundation

Agrobacterium   A soil bacterium that can transfer its own genes across the plant cell wall and membrane into the nucleus. The bacterium's DNA integrates into the plant's genome. The use of *Agrobacterium* allows the introduction of genes from any species to be engineered into crops.

Alleles   One of two or more alternate forms of a gene occurring at the same position on a chromosome. A gene for eye color, for example, may exist in two allelic forms, one for blue and one for brown.

Allergen   A substance that when eaten or inhaled causes an allergic reaction. Common allergens are dust, pollen, and pet dander.

Allergenicity   The tendency to cause an allergic reaction.

Amino acids   Organic molecules that are the building blocks of proteins. There are at least 20 different kinds of amino acids in living things. Proteins are composed of different combinations of amino acids assembled in chain-like molecules.

Amniocentesis   A diagnostic test during pregnancy to check for chromosomal abnormalities in the baby.

Amylose   A component of starch.

Antibiotic   A drug that kills or slows the growth of bacteria.

Antibiotic resistance   The property of an organism to be resistant to antibiotic treatment.

Antioxidants   Chemical compounds or substances thought to protect body cells from oxidation, which has damaging effects.

APHIS   The USDA's Animal and Plant Health Inspection Service.

Arabidopsis   A flowering plant related to cabbage and mustard that is so small that it can be grown on a Petri dish, and the first plant to have its entire genome sequenced.

Aspergillus   A common mold that causes food spoilage. Some species can infect damaged corn kernels and produce mycotoxins that can cause cancer.

ATP (adenosine triphosphate)   Supplies large amounts of chemical energy to cells for biochemical processes.

Atrazine   One of the most widely used agricultural herbicides. Atrazine persists in ground water and can induce hermaphroditism in frogs.

Bacillus thuringiensis (Bt)   A group of soil bacteria found worldwide, which produce a class of proteins highly toxic to the larvae (immature forms) of certain taxonomic groups of insects. Bacterial spores containing the toxin are used by organic farmers as an environmentally benign, commercial pesticide. The Bt toxin will kill earworms, budworms, gypsy moth larvae, Japanese beetles, and other insect pests. Since 1989, genes expressing the cry proteins have been introduced into plants to confer insect resistance. Bt also refers to the insecticidal toxins.

Bacteria   Single-celled organisms.

Biodiversity (biological diveriisty)   The total variability within and among species of living organisms and their habitats. Biodiversity is divided into three levels: genetic (genes within a local population or species), taxonomic (the species comprising all or part of a local community), and ecological (the communities that compose the living parts of ecosystems).

Biotic Controls   Biological solutions to insect pest problems. For example, application of beneficial insects to control pest populations.

Bt crop   A crop plant genetically engineered to produce insecticidal toxins derived from the bacterium *Bacillus thuringiensis*. Current commercial Bt crops include Bt cotton, Bt corn, and Bt soybeans.

CAMBIA (Center for the Application of Molecular Biology to International Agriculture)   A nonprofit research institute.

Carbamate pesticides   A salt or ester of carbamic acid used as a pesticide and known to cause death of animals at high concentrations. Aldicarb is effective against thrips, aphids, spider mites, lygus, fleahoppers, and leafminers but is mainly used against nematodes. Carbofuran is the most toxic of the carbamate pesticides used to control insects in potatoes, corn, and soybeans. Carbaryl kills beneficial insect and crustacean species along with the target pests. 2-(1-Methylpropyl)phenyl methylcarbamate is used as an insecticide on rice and cotton.

Carcinogenic   Any substance or agent that tends to cause cancer.

Carbohydrates   A class of organic molecules that include sugar and starches.

Carotenoid   Precursor to Vitamin A. A class of yellow to red pigments found widely in plants and animals.

Cell   The basic functional unit of an organism, usually with a nucleus, cytoplasm, and an enclosing membrane. All plants and animals are composed of one or more microscopic cells.

Chemical fertilizers   Synthetically produced mineral forms of nitrogen, phosphorous, and potassium given to plants to enhance growth. Require large amounts of fossil fuels for synthesis.

Chimera   An individual composed of genetically distinct individuals.

Crohn's disease   A disease of the intestine, specifically the distal portion of the ileum, characterized by abdominal pain, ulceration, and fibrous tissue buildup.

Chromosomes   In bacteria, a circular strand of DNA that contains the hereditary information. In eukaryotic cells (higher organisms such as plants), chromosomes consists of linear strands of DNA comprised of tens of thousands of genes. They are found in the nucleus of every cell.

Compost   A 20:1 mixture of carbonaceous and nitrogenous organic matter that, in the presence of water and various microbes decomposes, into hums and aporphous organic particles. Used as a fertilizer on organic farms.

Conventional farming   Conventional agriculture. A catch-all term used to describe diverse farming methods. At one end of the continuum are farmers who use synthetic pesticides and fertilizers to maximize short-term yields. At the other end are growers who use chemicals sparingly and embrace the goals of ecological farming. Increasingly, many conventional farmers, particularly in the United States, are growing GE crops.

Cover crops   Plants that are grown to be turned back into the soil for nutrients and organic matter. Two examples of cover crops are vetch and bell beans.

Crop yield   Agricultural output. A measure of yield per unit area of land.

Cultivar   A group of individual plants within a species that are uniform in appearance. Progeny of a particular cultivar share the same attributes of the parent plant.

DHMO   Dihydrogen monoxide. Water.

Diuron   An herbicide used as a weed killer. Persists in ground water and highly toxic to aquatic invertebrates.

DNA (deoxyribonucleic acid)   The basic genetic material found in all living cells (and some viruses), providing the blueprint (i.e. genes) for construction of proteins. DNA is composed of sugars, phosphates, and bases arranged in a "double helix," a double stranded, chain-like molecules composed of nucleotide base pairs.

Domestication   An artificial selection process to produce plants that have fewer undesirable traits.

Ecological farming   A farming system that involves the coordination of various elements such as crop rotation, variety selection, fertilization, tillage, plant protection, productivity, crop quality, and environmental compatibility for growing particular crops.

Ecosystem   A collection of organisms interacting with the surrounding physical environment resulting in a functioning ecological unit.

Enzymes   Proteins that cause or regulate specific chemical reactions in the cell.

Fitness   A relative measure of an organism's reproductive efficiency (i.e., the relative probability of reproduction of a genotype), generally referring to Darwinian fitness. Components of fitness include survival, rate of development, mating success, and fertility, and pathogenicity in the case of microbes. Fitness is germane to hazard assessment of organisms engineered to contain foreign genes. Also called adaptive value.

Fluoridation   Addition of fluoride to community drinking water.

Frostban   A genetically modified bacteria used to protect plants from frost.

Gamete   A reproductive cell such as ovule or pollen. Female and male gametes unite to form a single cell called the zygote, which, through division, generates an embryo and ultimately a progeny individual.

Gene   Genes are responsible for hereditary characteristics of plants and animals. Genes occur at specific locations on a chromosome. A gene is a sequence of DNA bases. Some genes direct the synthesis of one or more proteins, while others have regulatory functions (controlling the expression of other genes). A gene may be made up of hundreds of thousands of DNA bases but are typically composed of one to a few thousand.

Gene (genetic) marker (or marker gene)   A DNA sequence, gene, or trait that is used to track a genetic event. A selectable marker gene produces an identifiable phenotype (i.e., observable characteristics) that can be used to track the presence or absence of other genes (e.g., genes of interest) on the same piece of DNA transferred into a cell.

Gene flow   The movement of genes from one population to another by way of hybridization of related and sexually compatible individuals in the two populations. In plants, gene flow takes place by transfer of pollen (male gametes) or seeds.

Gene silencing   The interruption or suppression of the expression of a gene at transcriptional or translational levels. A gene can be silenced by genetic engineering.

Genetic engineering   The alteration of an organism's genome by introducing, modifying, or eliminating specific genes using transformation rather than conventional breeding methods. Differs from older methods of genetic modification in that a gene from any species can be inserted into an organism.

Genetic modification   The alteration of an organism's genome by human intervention, by introducing, modifying, or eliminating specific genes. Methods include conventional plant breeding, such as pollen transfer, embryo rescue, grafting, and mutagenesis. Genetic modification usually is restricted to gene transfer within a species but interspecies mixing can also be achieved.

Genetic resource   Genetic material serving as a resource for human use. For plants, this includes modern cultivars (varieties), landraces, and wild and weedy relatives of crop species and the genes that these plants contain. Plant breeders and genetic engineers rely on a broad, diverse genetic base to enhance crop yields, quality, or adaptation to environmental extremes.

Genetics   The study of gene structure and action and the patterns of inheritance of traits from parents to offspring. Genetics is the branch of science that deals with the inheritance of biological characteristics.

Genetic transformation   The process where from a donor organism is transferred directly into a recipient cell using *Agrobacterium* or mechanical methods to produce a transgenic organism.

Genome   An organism's total genetic content. The entire hereditary material of a cell or a virus, including the full complement of functional and nonfunctional genes. In higher organisms (including plants, animals, and humans) the genome comprises the entire set of chromosomes found within the cell nucleus.

Genomics   The scientific field of study that seeks to understand the nature (organization) and specific function of genes in living organisms.

Germplasm   The total genetic variability available to a particular population of organisms, represented by the pool of germ cells (sex cells, the sperm or egg). Also used to describe the plants, seeds, or other plant parts useful in crop breeding, research, and conservation efforts, when they are maintained for the purpose of studying, managing, or using the genetic information they possess.

Glassy-winged Sharpshooter   An insect that transmits the *Xylella fastidosa* bacterium to grape vines.

Glyphosate   An herbicide that targets a plant metabolic process and is therefore not toxic to animals. Does not persist in ground water.

GRRF   Genetics Resources Recognition Fund.

Heirloom plant   An older open-pollinated variety that is no longer used in modern large-scale agriculture. Often selected by an individual and then passed down from generation to generation.

Herbicides   A chemical used to kill weeds.

Herbicide-tolerant (HT) crop   A crop able to survive the application of one or more synthetic chemical herbicides, many of which are toxic to both crops and weeds. Includes those conventionally bred and those genetically engineered to contain genes (or mutated genes) that make them insensitive to or able to detoxify herbicides.

Homeopathic   The application of treating and preventing disease with minute doses of drugs or remedies to enhance the organism's natural defense mechanisms.

Hormones   Substances (usually proteins) that influence chemical reactions and regulate various cellular functions.

Hybrid   The offspring from plants of the same species but different varieties.

Hybrid vigor   Offspring from parents of "inbred" parent lines result in higher yield.

Hybridization   The production of offspring (hybrids) from genetically unlike parents, by natural processes or by human intervention (i.e., artificial selection). In plant breeding, includes the process of cross-breeding two different varieties to produce hybrid plants. If the hybrid is more fit than either parent; the condition is called hybrid vigor (or heterosis). Hybrid offspring may result from gene flow between domesticated crops and wild relatives.

Inbred   A self-pollinated plant that is genetically uniform.

Insect pheromones   Chemical substances that help insects communicate with each other.

Insecticide   A chemical used to kill insects.

Intellectual Property (IP)   A product of the intellect that has commercial value, including patents on seeds, inventive methods, or gene sequences.

Invertebrates   An animal lacking a backbone, such as insects and snails.

IRRI   International Rice Research Institute.

Landrace   Refers to the particular kinds of plant varieties that are farmer-selected in areas where local subsistence agriculture is practiced. The term is usually applied to plant varieties that were domesticated by farmers, and further modified by native and also immigrant farmers. Landraces are highly adapted to particular soil types and microclimates in specific locales. Landrace have a broad genetic base (highly heterozygous) resulting from centuries of development and adaptation. Landraces are an important source of diverse genes for plant breeders and geneticists.

Leptidoptera   A class of insects: moths and butterflies.

Ligases   Enzymes that initiate the linkage of two segments of DNA.

Marker   See genetic marker

Marker-assisted breeding   The use of genetic fingerprinting techniques to introduce genes of interest from one plant variety to another. Relies on knowledge of DNA sequences in a particular genomic region.

Messenger RNA (mRNA)   The form of RNA that carries a copy of a specific sequence of genetic information (a gene) from the DNA in the cell nucleus to the ribosomes in the cytoplasm, where it is translated into a protein.

Metachlor   Herbicide used in soybean fields to control weeds. Metachlor is a known groundwater contaminant and is included in a class of herbicides that are suspected to be toxic.

Monoculture   The practice of planting the same genetically uniform crop year after year. Monoculture can lead to higher yields (because planting, pest control and harvesting can be standardized) but also to large-scale crop failure if the crop becomes susceptible to a disease.

Monounsaturated fatty acids   Long-chained molecules found in nuts, avocados, olive oil, grapeseed oil, peanut oil, flaxseed oil, sesame oil, corn oil, and canola oil.

Mutagen   An agent that can cause a mutation. Various kinds of chemicals, viruses, radiation and sun exposure have been shown to be mutagenic.

Mutation   An alteration of genetic material such that a new variation is produced.

Mutation breeding   Seeds are put in a highly carcinogenic solution or treated with radiation to induce random changes in the DNA. After germination, surviving seedlings that have new and useful traits are then used by breeders.

Nematodes   Plant-parasitic roundworms which attack roots and underground parts of plants.

NIH (National Institutes of Health)   A federal agency that conducts and funds biomedical research.

Nitrogen   An element that occupies 78% of our atmosphere. It is a part of all living tissues and amino acids. Nitrogenous formulations are used as fertilizer.

Non-target effect   An effect stemming from intentional introduction of plants, chemicals, proteins, or microbes to natural, agronomic, or forest ecosystems. (E.g. The highly specific *Bacillus thuringiensis* (Bt) toxin is meant to target pests of a particular crop such as earworms, budworms, gypsy moth larvae, Japanese beetles, and other insect pests but non-target effects on other insects has also been documented).

Nucleotide   The basic building block of a nucleic acid. It consists of any one of four specific purine or pyrimidine bases attached to a ribose or deoxyribose sugar and phosphate group.

Nucleus   A structure (organelle) found in all eukaryotic cells. It contains the chromosomes and is enclosed by a nuclear membrane.

Open pollination (OP)   Natural pollination via wind, insects, etc. without the use of hybrids.

Organophosphate insecticide metabolites   Breakdown products of organophosphate insecticides. Have been found in urine of adult farmworkers and children exposed to the organophosphate residues.

Outbreeding (outcrossing)   Sexual combination between distantly related members of the same species, in contrast to inbreeding, mating between closely related members. In outbreeding plants, pollen and egg come from plants that are genetically different, permitting gene flow between varieties.

Pest   Any species that interferes with human activities, property, or health, or is otherwise objectionable. Economically important pests of agricultural crops include weeds, arthropods (including insects and mites), microbial plant pathogens, and nematodes (roundworms), as well as higher animals (e.g., mammals and birds).

Pesticide   Any substance or agent employed to destroy a pest organism. Common pesticides include insecticides (to kill insects), herbicides (to kill weeds), fungicides to kill (fungi), and nematicides (to kill nematodes).

Patent   Granted to an inventor; it protects his/her exclusive right to the invention.

Perennials   Plants that live for more than 1 or 2 seasons. They do not die after producing seeds.

Pesticides   Chemicals used to control pests.

Photosynthesis   The process by which plants convert sunlight to chemical energy.

Photovoltaic cells   Solar technology that converts the sunlight energy into electricity.

PIPRA   Public-sector Intellectual Property Resource for Agriculture.

Plant breeding   Manipulation of plant species in order to create desired genetic modifications for specific purposes.

Plumpox (PPV) virus   A viral disease of stone fruits.

Plutella xylostella   Diamondback moth.

Pollination   The transfer of pollen (male gamete) to the plant carpel, where the female gamete resides. This can be done by insects, birds, wind, water, or humans.

Polyunsaturated fatty acids   Important for maintaining membranes of cells, inflammation regulation process, blood clotting regulation process, and the absorption of vitamins A, D, E, and K.

Potassium   An element needed for plant growth and used as a fertilizer.

Protein   Encoded by genes, proteins are composed of amino acids arranged in precise sequences and joined by peptide linkages. Proteins can serve as enzymes, regulators of gene activity, transporters, hormones, or other catalytic and structural elements.

Prunus domesticus   Latin name for the plum tree.

Psoralens   Toxic compounds produced by celery that discourage predators.

PVP   Plant Variety Protection Act.

Recombination   The process by which alleles are exchanged between pairs of chromosomes (those inherited from the maternal and paternal parents) during sexual reproduction. Recombination creates new combinations of alleles at different loci along the chromosome.

Rennet   Also known as rennin and chymosin. Enzyme that is used for curdling milk and making cheese.

Restriction enzymes   Enzymes that cut DNA molecules as specific base pairs.

RNA (ribonucleic acid)   A single-stranded genetic material critical for protein synthesis in living cells.

Rootone   A plant hormone that induces rooting.

Rotenone   An insecticide extracted from plants that is also toxic to humans.

Roundup   The brand name for the glyphosate herbicide produced by Monsanto used to control weeds.

Salmonella enteritidis   A bacterium that causes food poisoning in humans. It is found on raw eggs and poultry.

Scion   Top part of the tree or a shoot with buds that is used for grafting.

Seed stock   The seed, tubers, and roots saved by a farmer after each harvest to be used for the next crop production. Seed supply.

Snomax   A strain of bacteria related to Frostban that promotes freezing. Snomax is being used in ski resorts on snow-making machines.

Soil erosion   A process whereby wind and water carry away soil, depleting the amount of soil available. Human activity, such as overcultivation and compaction, can also lead to soil erosion.

Soil fumigants   Pesticides used to fumigate soil to prepare for planting, to control weeds, and kill nematodes.

Soil solarization   An effective, nonchemical approach to controlling weeds and soil pest problems. The technique involves covering moist soil with a thin, clear plastic for six weeks in the heat of the summer.

Solanine   Toxic bitter chemical produced by green potatoes, eggplants, tomatoes, and peppers as a natural defense mechanism. Exposure to grocery store light causes the potato solanine levels to increase.

Species   A taxonomic category of life forms, usually consisting of organisms that are sexually compatible and may actually or potentially interbreed in nature. The scientific (or Latin) name of a species includes the genus name and species designation, with the genus placed first (e.g., *Bacillus thuringiensis*).

Spontaneous mutation   A mutation that occurs spontaneously as opposed to one that is induced by chemicals or radiation.

Steinernema feltiae A beneficial nematode that attacks the larvae of soil and above-ground insect pests such as fungus gnat, various flies, flea beetles, and some plant parasitic nematodes.

Submergence tolerance A trait that allows young rice plants to withstand or tolerate 1–2 weeks of submergence.

Surfactant A substance that reduces surface tension of the liquid used to dissolve it and increase the solubility of organic compounds. Found in herbicides.

Symbiotic relationship An ecological relationship between organisms of two or more different species benefiting both species.

Synthetic fertilizers Fertilizers made from fossil fuels. Examples of synthetic fertilizers are ammonia, ammonium sulfate, and urea.

Teosinte *Zea mexicana*, wild ancestral corn from Mexico and Central America. The seeds are not united on a cob. Rather, the female inflorescence (the ear) consists of a single row of six or more seeds, each of which contains a hard, flinty endosperm, like popcorn, covered by a tough shell.

Transgene Genes that are inserted into the genome of a cell via genetic transformation (genetic engineering). Along with the genes of interest, a transgene may contain promoter, other regulatory, and marker genetic material.

Transgenic plant A plant containing transgenes. The transgenes are passed onto the offspring.

Transposable elements Pieces of DNA that move around. They insert themselves into new places and sometimes pick up pieces of other genes.

Trichogramma Extremely tiny wasps that are beneficial insects. They parasitize the eggs of moths and butterflies.

USDA (United States Department of Agriculture) A Federal agency that develops and administer agricultural policies and programs

VAD Vitamin A deficiency.

Variety A category used in the classification of plants and animals below the species level. A variety consists of a group of individuals that differ distinctly from but can interbreed with other varieties of the same species. The characteristics of a variety are genetically inherited.

Vector A circular, nonchromosomal DNA found in bacteria (called plasmid) that can self-replicate and is used to carry new genes into cells. In plant pathology, a vector is an organism capable of transmitting a pathogen from one host to another, such as plant-feeding insects that transmit viruses.

Weed Any unwanted plant that interferes with human activities (including farms and gardens) or natural habitats. Plants may be considered weeds for diverse reasons (e.g., rapid growth, persistence, invasiveness, toxicity to livestock).

Vermiculite Silicate minerals used for heat insulation, plaster, packing material, and as planting medium.

Vital-force theory   The theory that a vital force determined the difference between organic and inorganic compounds. Organic materials isolated from plants and animals were thought to contain a vital force, while inorganic materials did not.

Vitamin A   A fat-soluble compound found in fish-liver oils, milk, green and yellow vegetables, and egg yolk. It is required for cell growth and development, epithelial tissue growth and protection, and normal vision.

Vitamin D   A fat-soluble compound found in milk and fish-liver oils that is required for tooth and bone growth.

Warrior   A pyrethroid insecticide used to kill weevils and control maggots and flies.

Waxy   A gene that encodes an enzyme for amylose synthesis.

Xa21   A rice gene that confers resistance to most strains of *Xanthomonas oryzae pv. oryzae.*

Xanthomonas oryzae   A bacteria that causes bacterial blight of rice.

Xylella fastidosa   A tiny bacterium that causes plant diseases that are economically important, such as Pierce's disease, a lethal disease to grape vines.

References

Preface

Conway, Gordon. 1997. *The Doubly Green Revolution*. Ithaca, NY: Cornell University Press.

National Organic Program Standards. http://www.ams.usda.gov/nop/NOP/standards.html (accessed June 7, 2006).

United Nations, Department of Economic and Social Affairs, Population Division. 2007. *World Population Prospects: The 2006 Revision*. New York: United Nations.

United States Department of Agriculture Food and Nutrition Service. http://www.fns.usda. gov/fns/food_safety.htm (accessed June 7, 2006).

Chapter 1

Catling, H. D. 1992. *Rice in Deep Water*. Manila, Philippines: International Rice Research Institute.

Dey, M. M., and H. K. Upadhyaya. 1996. Yield loss due to drought, cold and submergence tolerance. In R. E. Evenson, R. W. Herdt, and M. Hossain, (eds.). *Rice Research in Asia: Progress and Priorities*. UK: International Rice Research Institute in collaboration with CAB International.

Freeman, D. 1970. *Report on the Iban*. London: Athlone.

Hamilton, R.W. 2003. *The Art of Rice, Spirit and Sustenance in Asia*. Los Angeles: UCLA Fowler Museum of Cultural History.

Herdt, R. W. 1991. Research priorities for rice biotechnology. In G. S. Khush and G. H. Toenniessen (eds.). *Rice Biotechnology*. Wallingford, UK: CAB International.

Huke, R. E., and E. H. Huke. 1990. *Rice: Then and Now*. Manila, Philippines: International Rice Research Institute.

International Rice Genome Sequencing Project (IRGSP). 2005. The map-based sequence of the rice genome. *Nature* 436: 793–800.

Xu K., X. Xu, P. C. Ronald, and D. J. Mackill. 2000. A high-resolution linkage map in the vicinity of the rice submergence tolerance locus Sub1. *MGG* 263: 681–689.

Xu, Kenong, Xia Xu, Takeshi Fukao, Patrick Canlas, Sigrid Heuer, Julia Bailey-Serres, Abdel Ismail, P. C. Ronald, and David J. Mackill. 2006. Sub1A encodes an ethylene responsive-like factor that confers submergence tolerance to rice. *Nature* 442: 705–708.

CHAPTER 2

Aroian Lab. 2006. *Bacillus thuringiensis*: History of Bt. University of California–San Diego. http://www.bt.ucsd.edu/bt_history.html (accessed August 22, 2006).

Asami, Danny K., Yun-Jeong Hong, Diane M. Barrett, and Alyson E. Mitchell. 2003. Comparison of the total phenolic and ascorbic acid content of freeze-dried and air-dried marionberry, strawberry, and corn grown using conventional, organic, and sustainable agricultural practices. *Journal of Agriculture and Food Chemistry* 51: 1237–1241.

Ascherio, Alberto, Honglei Chen, Marc G. Weisskopf, Eilis O'Reilly, Majorie L. McCullough, Eugenia E. Calle, Michael A. Schwarzschild, and Michael J. Thun. 2006. Pesticide exposure and risk for Parkinson's disease. *Annals of Neurology* 60: 197–203.

Baker, Brian, Charles Benbrook, Edward Groth III, and Karen Lutz Benbrook. 2002. Pesticide residues in conventional, IPM-grown, and organic foods: Insights from three U.S. data sets. *Food Additives and Contaminants* 19(5): 427–446.

Bellinger, R. G. 1996. Pest Resistance to Pesticides. Clemson University: Pesticide Information Program. http://entweb.clemson.edu/pesticid/Issues/pestrest.pdf (accessed June 15, 2006).

Blank, Stephen. 1998. *The End of Agriculture in the American Portfolio*. Westport, CT: Quorum Books.

CBS News. 2006. Farmers pay price as fuel costs soar. May 6.

Curl, Cynthia L., Richard A. Fenske, and Kai Elgethun. 2003. Organophosphorus pesticide exposure of urban and suburban preschool children with organic and conventional diets. *Environmental Health Perspectives* 111(3): 377–382.

Denison, R. Ford, Dennis Bryant, and Thomas E. Kearney. 2004. Crop yields over the first nine years of LTRAS, a long-term comparison of field crop systems in a Mediterranean climate. *Field Crops Research* 86: 267–277.

Department of Pesticide Regulation. 2006. DPR releases 2004 pesticide use data; more nature-friendly chemicals gain favor. State of California News Release, January 24. http://www.cdpr.ca.gov/docs/pressrls/2006/060124.htm (accessed June 15, 2006).

Diver, Steve, George Kuepper, and Preston Sullivan. 2001. *Organic Sweet Corn Production: Horticulture Production Guide*. Insect Pest Management. ATTRA—National Sustainable Agriculture Information Service. Fayetteville, AR, September. http://attra.ncat.org/attra-pub/sweetcorn.html#insect (accessed August 28, 2006).

Elmore, C., et al. 1997. *Soil Solarization. A Nonpesticidal Method for Controlling Diseases, Nematodes, and Weeds*. University of California, DANR Publication 21377.

Gewin, Virginia. 2004. Organic FAQs: Can organic farming replace conventional agriculture? *Nature* 428 (April 22): 798.

Goldstein, Nora. 2005. Source Separated MSW Composting in the U.S.: From Michigan and Minnesota to Washington and California, household organics are being separated and sent to specially-designed composting sites. Part II. *Biocycle* 46(12): 20.

Guthman, Julie. 2004. The trouble with 'organic lite' in California: A rejoinder to the 'conventionalisation' debate. *Sociologia Ruralis* 44(3): 301–316.

Hoitink, H. A. J., and M. E. Grebus. 1994. Stats of biological control of plant diseases with composts. *Compost Science and Utilization* 2(2): 6–12.

King, F. H. 1911. *Farmers of Forty Centuries: Farmers of China, Korea, and Japan*. Madison, WI: Mrs. F. H. King.

Klein, Jill. 1989. *The Use of Green Manures in Field and Vegetable Crop Systems*. Annual Report to the California Energy Commission. From: The development of a cover crop program for the central valley. Master's thesis, UC Davis, Davis, California.

Kramer, Sasha B., John P. Reganold, Jerry D. Glover, Brendan J. M. Bohannan, and Harold A. Mooney. 2006. Reduced nitrate leaching and enhanced denitrifier activity and efficiency in organically fertilized soils. *Proceedings of the National Academy of Sciences USA* 103(12): 4522–4527.

Legarre, Paola, E. Nakata, and G. Peer. 2001. Pricing strategy analysis: Farm and community supported agriculture. M.B.A. project paper. Leavey School of Business, Santa Clara, California.

Liebhardt, Bill. 2001. Get the facts straight: Organic agriculture yields are good. Information Bulletin. *Organic Farming Research Foundation* 10(Summer): 1, 4–5.

Lumsden, R. D., J. A. Lewis, and P. D. Millner. 1983. Effect of compost sewage sludge on several soilborne pathogens and disease. *Phytopathology* 73(11): 1543–1548.

Maeder, Paul, Andreas Fliessbach, David Dubois, Lucie Gunst, Padruot Fried, and Urs Niggli. 2002. Soil fertility and biodiversity in organic farming. *Science* 296 (May 31): 1694–1697.

National Institute of Environmental Health Sciences (NIEHS). 2003, May 1. Agricultural pesticide use may be associated with increased risk of prostate cancer. U.S. Department of Health and Human Services: NIH News. http://www.niehs.nih.gov/oc/news/aghltst.htm (accessed August 22, 2006).

Natural Resources Conservation Service (NRCS). 2003, July. Soil Erosion. Natural Resources Inventory Annual, 2001. http://www.nrcs.usda.gov/TECHNICAL/land/nri01/nri01eros.html (accessed August 22, 2006).

Natural Resources Conservation Service. 2006. http://www.nrcs.usda.gov/

Pimentel, D., H. Acquay, M. Biltonen, P. Rice, M. Silva, J. Nelson, V. Lipner, S. Giordano, A. Horowitz, and M. D'Amore. 1993. Assessment of environmental and economic impacts of pesticide use. In D. Pimentel and H. Lehmann (eds.). *The Pesticides Questio: Environment, Economics and Ethics*. New York/London: Chapman and Hall, pp. 47–84.

Pimentel, David, Paul Hepperly, James Hanson, David Douds, and Rita Seidel. 2005. Environmental, energetic, and economic comparisons or organic and conventional farming systems. *BioScience* 55(7): 573–582.

Pollan, Michael. 2006. *The Omnivore's Dilemma: A Natural History of Four Meals*. New York: Penguin Press.

PUR (Department of Pesticide Regulation Pesticide Use Reporting). 2004a. The top five pesticides used in each county in 2004 and the top five sites of use for each of these pesticides. State of California. http://www.cdpr.ca.gov/docs/pur/pur04rep/top5_ais.pdf (accessed May 26, 2006).

PUR (Department of Pesticide Regulation Pesticide Use Reporting). 2004b. Pounds of active ingredient by county. State of California. http://www.cdpr.ca.gov/docs/pur/pur04rep/lbsby_co.pdf (accessed April 17, 2006).

Shapouri, Hosein, James Duffield, Andrew McAloon, and Michael Wang. 2004. *The 2001 Net Energy Balance of Corn-Ethanol*. U.S. Department of Agriculture: Office of the Chief Economist (OCE), Agricultural Research Service (ARS). Washington, D.C.: U.S. Department of Energy. http://www.usda.gov/oce///reports/energy/net_energy_balance.pdf

Texas A&M University. 2005, April 6. "Dead zone" arrives early in Gulf in 2005. Texas A&M University News and Information. http://communications.tamu.edu/newsarchives/05/040605–14.html (accessed August 22, 2006).

US EPA. 1996. *Organic Strawberry Production as an Alternative to Methyl Bromide.* EPA 430-R-96–012.

Van Horn, Mark. 1995. *Compost Production and Utilization: A Grower's Guide.* UCDANR Pub. 21514.

Warner, Melanie. 2006. Wal-Mart eyes organic foods, and brand names get in line. *New York Times,* May 12.

Weeks, Carly. 2006, April 17. Food Inc. Swallows Organics: Multinationals Move to Corner Niche Market. CanWest News Service. http://www.canada.com/edmontonjournal/news/story.html?id=15471e4c-6c19–452d-af5e-750d61a01888

Zahm S. H., and A. Blair. 1992. Pesticides and non-Hodgkins lymphoma. *Cancer Research* 52(19): 5485s–5488s.

CHAPTER 3

Agraquest. Products. http://www.agraquest.com/products/index.html

Appropriate Technology Transfer for Rural Areas (ATTRA). 2005. *Reduced-Risk Pest Control Factsheet: Insect IPM in Apples: Kaolin Clay.* ATTRA Publication. (Last modified May 11). http://attra.ncat.org/attra-pub/PDF/kaolin-clay-apples.pdf (accessed August 24, 2006).

Aroian Lab. 2006. *Bacillus thuringiensis*: History of Bt. University of California San Diego. http:// www.bt.ucsd.edu/bt_history.html (accessed August 24, 2006).

Carson, Rachel. 1962. *Silent Spring.* Boston: Houghton Mifflin.

Chrispeels, Maarten J., and David E. Sadava. 1994. *Plants, Genes, and Agriculture.* Boston: Jones and Bartlett Publishers.

Diver, Steve. 1999. *Biodynamic Farming and Compost Preparation.* National Sustainable Agriculture Information Service. ATTRA Publication #IP137.

Hegege, Maya (interviewer), and Jarrell, Randall (ed.) 2003. *The Early History of USC's Farm and Garden: Interviews with Paul Lee, Phyllis Norris, Orin Martin, Dennis Tamura.* Santa Cruz: University of Calilfornia–Santa Cruz, University Library. http://library.ucsc.edu/reg-hist/farmgarden.html

Howard, Albert. 1940. *An Agricultural Testament.* London: Oxford University Press.

Howard, Albert. 1947. *The Soil and Health: A Study of Organic Agriculture.* New York: The Devin-Adair Company.

Kupfer, David. 2001. The Organic Farming Movement: Trailblazers, Heroes, and Pioneers. http://www.wildnesswithin.com/kupfer.html

Madden, Patrick. 1990. The Early Years of the LISA, SARE, and ACE Programs. http://wsare.usu.edu

National Research Council, Board on Agriculture. 1989. *Alternative Agriculture.* Committee on the Role of Alternative Farming Methods in Modern Production Agriculture. Washington D.C.: National Academy Press.

Pimentel, D., Hepperly, P., Hanson, J., Seidel, R., and Douds, D. 2005. Environmental, energetic, and economic comparisons of organic and conventional farming systems. *Bioscience* 55(7): 573–582.

Rodale, J. I. 1976. *How to Grow Vegetables and Fruits by the Organic Method*. Emmaus, PA: Rodale Books.

Sooby, Jane. 2006. *Investing in Organic Knowledge*. OFRF. Santa Cruz, CA.

Sustainable Agriculture Research and Education. http://www.sare.org/projects/

Trewavas, Anthony. 2006. Urban myths of organic farming. *Nature* 410: 410.

United States Department of Agriculture (USDA), Agriculture Marketing Service. National Organic Program. http://www.ams.usda.gov/nop/indexNet.htm

United States Department of Agriculture (USDA). 1980, July. *Report and Recommendations on Organic Farming*. GPO no. 620–220/3641. Washington, D.C.: USDA Study Team on Organic Farming.

United States General Accounting Office. 1990, February. *Alternative Agriculture: Federal Incentives and Farmers' Opinions*. GAO/PEMD-90–12. U.S. Washington, D.C.: General Accounting Office.

Wehner, T. C., and C. Barrett. 2005. Cucurbit Breeding Horticultural Science: History of Plant Breeding: Plant Breeding Methods. North Carolina State University. Designed by C. T. Glenn. http://cuke.hort.ncsu.edu/cucurbit/wehner/741/hs741hist.html (accessed August 24, 2006).

CHAPTER 4

Bodner Research Web. 2004, October 27. Structure and Nomenclature of Organic Compounds, What Is an Organic Compound? Chemistry Department, Purdue University. http://chemed.chem.purdue.edu/genchem/topicreview/bp/1organic/1org_frame.html (accessed March 6, 2006).

Bronowski, J. 1956. *Science and Human Values*. New York: Harper & Row.

Brooke, John Hedley. 1995. *Thinking about Matter: Studies in the History of Chemical Philosophy*. Aldershot, Hampshire, UK: Ashgate Publishing.

Brown, J. R. 2003. Ancient horizontal gene transfer. *Nature Reviews Genetics* 4(February 1): 121–156.

Clark, Robin (ed.). 1999. Overview GEO-2000: Global Environment Outlook. Chapter 2, The State of the Environment—North America: Marine and Coastal Areas. United Nations Environment Programme (UNEP). http://www.unep.org/Geo2000/english/ (accessed on May 4, 2006).

Dubcovsky, J., and Dvorak, J., 2007. Genome plasticity a key factor in the success of polyploidy wheat under domestication. *Science* 316:1862.

Hamilton, R.W. 2003. *The Art of Rice, Spirit and Sustenance in Asia*. Los Angeles: UCLA Fowler Museum of Cultural History.

Gonsalves, Dennis. 1998. Control of papaya ringspot virus in papaya: A case study. *Annual Review Phytopathology* 36:415–437. http://arjournals.annualreviews.org/doi/pdf/10.1146/annurev.phyto.36.1.415 (accessed May 9, 2006).

Guyotat, Régis. 2005. La justice relaxe des faucheurs volontaires en invoquant le 'danger imminent' des OGM. *Le Monde*, December 11.

Jiang, N., Z. Bao, X. Zhang, S. R. Eddy, and S. R. Wessler. 2004. Pack-MULE transposable elements mediate gene evolution in plants. *Nature* 431(September 30): 569–573.

Juniata College, Science in Motion. 2004. Berry Full of DNA. http://services.juniata.edu/ScienceInMotion/middle/labs/life/DNA_strawberry.doc.

Lane, Jo Ann. 1994 (adapted 2006). History of Genetics Timeline: 1994 Woodrow Wilson Collection. National Health Museum: Access Excellence: Activities Exchange. http://www .accessexcellence.org/AE/AEPC/WWC/1994/geneticstln.html (accessed June 16, 2006).

Lucca, P., X. Ye, and I. Potykus. 2001. Effective selection and regeneration of transgenic rice plants with mannose as selective agent. *Molecular Breeding* 7: 43–49.

McBride, J. M. 2003. Wöhler/Berzelius Letters about Urea (1828). Chemistry Department, Yale University. http://classes.yale.edu/chem125a/125/history99/4RadicalsTypes/UreaLetter1828. html (accessed March 6, 2006).

National Agricultural Statistics Service (NASS). 2002, July 11. Hawaii Papayas. Hawaii Agricultural Statistics Service: Hawaii Department of Agriculture. http://www.nass.usda. gov/hi/fruit/xpap0602.pdf (accessed on May 4, 2006).

National Diabetes Information Clearinghouse (NDIC). 2005. Total Prevalence of Diabetes in the United States, All Ages, 2005. http://diabetes.niddk.nih.gov/dm/pubs/ statistics/index. htm#7 (accessed March 3, 2006).

Schumacher, E. F. 1973. *Small Is Beautiful*. New York: Harper & Row.

Song, Wen-Yuan, Guo-Liang Wang, Li-Li Chen, Han-Suk Kim, Li-Ya Pi, Tom Holsten, J. Gardner, Bei Wand, Wen-Xue Zhai, Li-Huang Zhu, Claude Fauquet, and Pamela Ronald. 1995. A receptor kinase-like protein encoded by the rice disease resistance gene, Xa21. *Science* 270(5343): 1804–1806.

Toenniessen, Gary H., John C. O'Toole, and Joseph DeVries. 2003. Advances in plant biotechnology and its adoption in developing countries. *Current Opinion in Plant Biology* 6(2): 191–198.

Tripathi, Savarni, Jon Suzuki, and Dennis Gonsalves. 2006. Development of genetically engineered resistant papaya for *Papaya ringspot virus* in a timely manner—A comprehensive and successful approach. *Methods in Molecular Biology/Molecular Medicine*. Totowa, NJ: Humana Press, Inc.

University of California Division of Agricultural and Natural Resources Statewide Biotechnology Workgroup. 2007. http://ucbiotech.org/

Wright, Susan. 2001. Legitimating Genetic Engineering. Foundation for the Study of Independent Social Ideas, Inc., AG Biotech Info Net. http://www.biotech-info.net/ legitimating.html (accessed March 3, 2006).

CHAPTER 5

Bennett, R., U. Kambhampati, S. Morse, Y. Ismael. 2006. Farm level economic performance of genetically modified cotton in Maharashtra, India. *Review of Agricultural Economics* 28:59–71.

Bernstein, L., J. A. Bernstein, M. Miller, S. Tierzieva, D. I. Bernstein, Z. Lummus, M. K. Selgrade, D. L. Doerfler, and V. L. Seligy. 1999. Immune responses in farm workers after exposure to *Bacillus thuringiensis* pesticides. *Environmental Heath Perspectives* 107: 575–582.

California Certified Organic Farmers (CCOF). Genetic Engineering. http://www.ccof. org/ ge_pol_ad.php (accessed March 28, 2006).

California Department of Food and Agriculture (CDFA). 2003. *A Food Foresight Analysis of Agriculture Biotechnology: A Report to the Legislature*. Food Biotechnology Task Force, 3. http://www.cdfa.ca.gov/exec/pdfs/ag_biotech_report_03.pdf (accessed May 4, 2006).

California Department of Food and Agriculture (CDFA). 2004. *2004 County Organic Crop and Acreage Report*. http://www.cdfa.ca.gov/is/i&c/docs/2004CountyReport.pdf (accessed March 29, 2006).

Cline, Harry. 2005. Kern supervisors support biotechnology. *Western Farm Press*, May 7.

Chrispeels, M. J., and D. E. Sadava. 1994. *Plants, Genes, and Agriculture*. Boston: Jones and Bartlett.

Davis, Joshua. 2003. Come to LeBow. *Wired* 11, February 2. http://www.wired.com/ wired/ archive/11.02/smoking.html?pg=1&topic=&topic_set (accessed March 29, 2006).

Durkin, Patrick R. 2003, March 1. *Glyphosate—Human Health and Ecological Risk Assessment Final Report*. Syracuse Environmental Research Associates, Inc. USDA Forest Service: Forest Health Protection. http://www.fs.fed.us/foresthealth/pesticide/risk_assessments/04a03_glyphosate.pdf

Farrell, Alexander E., Richard J. Plevin, Brian T. Turner, Andrew D. Jones, Michael O'Hare, and Daniel M. Kammen. 2006. Ethanol can contribute to energy and environmental goals. *Science* 311(January 27): 506–508.

Fernandez-Cornejo, Jorge, and William D. McBride. 2002. *Adoption of Bioengineered Crops*. Agricultural Economic Report No. AER810.

Fernandez-Cornejo, Jorge, and Margriet Caswell, 2006. The First Decade of Genetically Engineered Crops in the United States. USDA Economic Research Service. Economic Information Bulletin, Number 11.

Glare, Travis R., and Maureen O'Callaghan. 2000. *Bacillus thuringiensis: Biology, Ecology and Safety*. Weinheim: Wiley-VCH.

GM Science Review Panel. 2003, July. *GM Science Review, First Report: An Open Review of the Science Relevant to GM Crops and Food Based on Interests and Concerns of the Public*. http://www.gmsciencedebate.org.uk/report/pdf/gmsci-repor-full.pdf (accessed April 27, 2006).

Hawks, Bill. 2004, December 21. Letter to the National Association of State Departments of Agriculture. Washington, D.C.: Department of Agriculture, Office of the Secretary, Marketing and Regulatory Programs.

Huang, J. K., R. F. Hu, S. Rozelle, and C. Pray. 2005. Insect-resistant GM rice in farmers' fields: Assessing productivity and health effects in China. *Science* 308:688–690.

Huang, J. K., S. Rozelle, C. Pray, and Q. Wang. 2002. Plant biotechnology in China. *Science* 295:674–677.

Kershen, Drew. 2006. Health and food safety: The benefits of Bt corn. *Food and Drug Law Journal* 61:197.

Madison, Deborah. 1997. *Vegetarian Cooking for Everyone*. New York: Broadway Books.

Marvier, Michelle, C McCreedy, J. Regetz, and P. Karieva. 2007. A Meta-analysis of effects of Bt cotton and maize on nontarget invertebrates. *Science* 316:1475.

National Research Council and Institute of Medicine of the National Academies (NAS). 2004. *Safety of Genetically Engineered Foods: Approaches to Assessing Unintended Health Effects*. Washington D.C.: The National Academy Press.

Opinion of the Scientific Panel on Genetically Modified Organisms on a request from the Commission related to the safety of foods and food ingredients derived from insect-protected genetically modified maize MON 863 and MON 863 x MON 810, for which a request for placing on the market was submitted under Article 4 of the Novel Food Regulation (EC) No 258/97 by *Monsanto. European Food Safety Journal* 2004 (50): 1–25.

PUR (Department of Pesticide Regulation, Pesticide Use Reporting). 2004. State of California. http://www.cdpr.ca.gov/docs/pur/pur04rep/lbsby_co.pdf (accessed April 17, 2006).

Purcell, A. *Xylella fastidiosa* University of California, Berkeley, College of Natural Resources. http://nature.berkeley.edu/xylella/north/manual_section_ii.pdf (accessed March 29, 2006).

Sonoma County, California. 2005, November 8. Measure M, Genetically Engineered Organism Nuisance Abatement Ordinance County of Sonoma. http://smartvoter.org/ 2005/11/08/ca/ sn/meas/M/ (accessed March 7, 2006).

Sustainable Agricultural Research and Education (SARE). 2002. "Zea-later!" Organic Corn Treatment Spells End to Wormy Ears. http://www.sare.org/highlights/2002/ sweet_corn .htm (accessed March 29, 2006).

Syngenta. 2004, May. Bt-11 Sweet Corn Update. http://www.syngenta.com/en/downloads/ Bt_sweet_corn_update_5–04_final.pdf (accessed March 28, 2006).

Tabashnik, B. E. 1994. Evolution of resistance to *Bacillus thuringiensis*. *Annual Review of Entomology* 39:47–79.

Tabashnik, B. E., Yves Carrière, Timothy J. Dennehy, Shai Morin, Mark S. Sisterson, Richard T. Roush, Anthony M. Shelton, and Jian-Zhou Zhao. Insect resistance to transgenic Bt crops: Lessons from the laboratory and field. *Journal of Economic Entomology* 96: 1031.

Toenniessen, Gary H., John C. O'Toole, and Joseph DeVries. 2003. Advances in plant biotechnology and its adoption in developing countries. *Current Opinion in Plant Biology* 6(2): 191–198.

Qaim, M., and D. Zilberman. 2003. Yield effects of genetically modified crops in developing countries. *Science* 299: 900–902.

Wang, S., D. Just, and P. Pinstrup-Anderson. 2006. Tarnishing silver bullets: Bt technology adoption, bounded rationality and the outbreak of secondary pest infestations in China. Paper presented at the American Agricultural Economics Association Annual Meeting in Long Beach, CA, July 25.

Wu, F. 2006. Mycotoxin reduction in Bt corn: potential economic, health, and regulatory impacts. *Transgenic Research* 15: 277–289.

CHAPTER 6

Ewen, S. W. B., and A. Pusztai. 1999. Effect of diets containing genetically modified potatoes expressing *Galanthus nivalis* lectin on rat small intestine. *The Lancet* 354:1353–1354.

Ho, Mae-Wan (ed.). 2002. Mice prefer non GM. *Institute of Science in Society*: ISIS 13/14 (February). http://www.isis.org.uk/MicePreferNonGM.php? (accessed March 31, 2006).

Holt, Jim. 2005. The way we live now: Madness about a method. *New York Times Magazine*, December 11. http://www.nytimes.com (accessed March 29, 2006).

Kennedy, Donald (ed.). 2006. Retraction of Hwang et al., Science 308 (5729) 1777–1783. Retraction of Hwang et al., Science 303 (5664) 1669–1674: Editorial Retraction. *Science* (*Letters*) 301(5759):335.

Losey, J. E., L. S. Rayor, M. E. Carter. 1999. Transgenic pollen harms monarch larvae. *Nature* 399: 214.

National Research Council and Institute of Medicine of the National Academies (NAS). 2006. Washington D.C. http://www.nasonline.org/site/PageServer?pagename=ABOUT_main_ page (accessed May 2, 2006).

Pew Initiative on Food and Biotechnology. 2002, May 30. Three Years Later: Genetically Engineered Corn and Monarch Butterfly Controversy. University of Richmond, The PEW Charitable Trusts. http://pewagbiotech.org/resources/issuebriefs/ monarch.pdf (accessed May 8, 2006).

Serageldin, Ismail. 2002. The rice genome: World poverty and hunger—The challenge for science. *Science* 296(5565): 54–58.

Smith, Jeffery M. 2004. *Seeds of Deception. Exposing Industry and Government Lies About the Safety of the Genetically Engineered Foods You Are Eating.* Portland: Chelsea Green Ltd.

CHAPTER 7

Ackermann-Liebrich, Ursula, Thomas Voegeli, Kathrin Gunter-Witt, Isabelle Kunz, Maja Zullig, Christian Schindler, Margrit Maurer, and Zurich Study Team. 1996. Home versus hospital deliveries: Follow up study of matched pairs for procedures and outcome. *BMJ* 313(November 23): 1313–1318.

Ahloowalia, B. S., M. Maluszynski, and K. Nichterlein. 2004. Global impact of mutation-derived varieties. *Euphytica* 135(2): 187–204.

Anderson, Alicia D., Jennifer M. Nelson, Shannon Rossiter, and Frederick J. Angulo. 2003. Public health consequences of use of antimicrobial agents in food animals in the United States. *Microbial Drug Resistance* 9(4): 373–379.

Catchpole, Gareth S., Manfred Beckmann, David P. Enot, Madhav Mondhe, Britta Zywicki, Janet Taylor, Nigel Hardy, Aileen Smith, Ross D. King, Douglas B. Kell, Oliver Fiehn, and John Draper. 2005. Hierarchical metabolomics demonstrates substantial compositional similarity between genetically modified and conventional potato crops. *Proceedings of the National Academy of Sciences USA* 102(40): 14458–14462.

Centers for Disease Control and Prevention. 2003. Mary Ann Liebert, Inc., Foodborne and Diarrheal Diseases Branch, Division of Bacterial and Mycotic Diseases. http://www.cdc.gov/narms/publications/2003/a_anderson_2003.pdf (accessed June 16, 2006).

Centers for Disease Control and Prevention (CDC). 2005, October 13. *Salmonella enteritidis.* National Center for Infectious Disease, Division of Bacterial and Mycotic Disease. Department of Health and Human Services. http://www.cdc.gov/ncidod/dbmd/diseaseinfo/salment_g.htm (accessed on May 5, 2006).

Cline, Austin (Guide). 2006. Evolution and the Law: Textbook Disclaimers. Should Students Be Told That Evolution Is 'Only' a Theory? About, Inc., a part of the New York Times Company. Centers for Disease Control and Prevention. http://atheism.about.com/library/FAQs/evo/blevo_law_disclaimers.htm (accessed June 5, 2006).

Committee on Drug Use in Food Animals, Panel on Animal Health, Food Safety, and Public Health, Board on Agriculture, National Research Council. 1999. *The Use of Drugs in Food Animals: Benefits and Risk.* Washington D.C.: National Academy Press.

Food and Drug Administration-Center for Veterinary Medicine. 2000, October 18 (revised January 5, 2001). The Human Health Impact of Fluoroquinolone Resistant Campylobacter Attributed to the Consumption of Chicken. http://www.fda.gov/cvm/Documents/RevisedRA.pdf (accessed June 16, 2006).

Fountain, H. 2006. On not wanting to know what hurts you. *New York Times*, January 15.

Gold, H. S., and R. C. Moellering, Jr. 1996. Antimicrobial-drug resistance. *New England Journal of Medicine* 335(November 7): 1445–1453.

Guyotat, Régis. 2005. La justice relaxe des faucheurs volontaires en invoquant le 'danger imminent' des OGM. *Le Monde*, December 11.

Humphrey, J. H., K. P. West, Jr., and A. Sommer. 1992. Vitamin A deficiency and attributable mortality among under-5-year-olds. *Bulletin of the World Health Organization* 70:225–232.

Kahn, Robert D. & Co. 2006. Frostban Field Trial. Client: Advanced Genetic Sciences 1987. http://www.rdkco.com/cs.tpl?action=show1&csid=1008553434135509 (accessed June 5, 2005).

Kidd, Riam S., Annette M. Rossignol, and Michael J. Gamroth. 2002. Salmonella and other *Enterobacteriaceae* in dairy-cow feed ingredients: antimicrobial resistance in western Oregon. *Journal of Environmental Health* 64(May): 9–16.

Ludwig, David S., Karen E. Peterson, and Steven L. Gortmaker. 2001. Relation between consumption of sugar-sweetened drinks and childhood obesity: a prospective, observational analysis. *Lancet* 357(February 17): 505–508.

Mazur, Allan. 2004. *True Warnings and False Alarms: Evaluating Fears about the Health Risks of Technology, 1948–1971*. Washington D.C.: Resources for the Future.

McNeil, Donald G. 2006. In raising the world's I.Q., the secret's in the salt. *New York Times*, December 16.

National Research Council and Institute of Medicine of the National Academies (NAS). 2004. *Safety of Genetically Engineered Foods: Approaches to Assessing Unintended Health Effects*. Washington D.C.: The National Academy Press.

Netherwood, Trudy, Susana M. Martín-Orúe, Anthony G. O'Donnell, Sally Gockling, Julia Graham, John C. Mathers, and Harry J. Gilbert. 2004. Assessing the survival of transgenic plant DNA in the human gastrointestinal tract. *Nature Biotechnology* 22:204–209.

Nordlee, J. A., S. L. Taylor, J. A. Townsend, L. A. Thomas, and R. K. Bush. 1996. Identification of a Brazil-nut allergen in transgenic soybeans. *New England Journal of Medicine* 334(March 14): 688–692.

Occupational Safety and Health Administration. 1998. Material Safety Data Sheet: Bonide Rotenone 5%. U.S. Department of Labor. http://www.biconet.com/botanicals/infosheets/rotenon5.pdf (accessed June 5, 2006).

Oklahoma Legislature (Representative Bill Graves). 2003. House Bill HB1504. http://www.lsb.state.ok.us/ (accessed June 5, 2006).

Olsen, Kenneth M., and Michael D. Purugganan. 2002. Molecular evidence on the origin and evolution of glutinous rice. *Genetics* 162: 941–950.

Owen, Sri. 1993/1994 (first U.S. edition). *The Rice Book: The Definitive Book on the Magic of Rice, with Hundreds of Exotic Recipes from Around the World*. New York: St Martin's Press.

Prescott, Vanessa E., Peter M. Campbell, Andrew Moore, Joerg Mattes, Marc E. Rothenberg, Paul S. Foster, T. J. V. Higgins, and Simon P. Hogan. 2005. Transgenic expression of bean α-amylase inhibitor in peas results in altered structure and immunogenicity. *Journal of Agricultural and Food Chemistry* 53(23): 9023–9030.

Rifkin, Jeremy, in collaboration with Nicanor Perlas. 1983/1984. *Algeny: A New Word—A New World*. New York: Viking/Penguin Books.

Ropeik David, and Gray, George. 2002. *Risk: a Practical Guide for Deciding What's Really Safe and What's Really Dangerous in the World Around You.* New York: Houghton Mifflin Company.

Sano, Y. (1984). Differential regulation of waxy gene expression in rice endosperm. *Theoretical and Applied Genetics* 68: 467.

Schubert, Rainer, Doris Renz, Birgit Schmitz, Walter Doerfler. 1997. Foreign (M13) DNA ingested by mice reaches peripheral leukocytes, spleen, and liver via the intestinal wall mucosa and can be covalently linked to mouse DNA. *PNAS* 94(3): 961–996.

Stein A. J., H. P. S. Sachdev, and Qaim Matin. 2006. Potential impact and cost-effectiveness of Golden Rice. *Nature Biotechnology*. 24: 1200–1201.

Wiegers, T. A., M. J. N. C. Keirse, J. van der Zee, and G. A. H. Berghs. 1996. Outcome of planned home and planned hospital births in low risk pregnancies: Prospective study in midwifery practices in the Netherlands. *BMJ* 313(November 23): 1309–1313.

Way, Tom. 2006 (revised). Dihydrogen Monoxide. Dihydrogen Monoxide Research Division— Dihydrogen Monoxide Info. http://www.dhmo.org/ (accessed June 2, 2006).

Wiley, Kip, Nick Vucinich, John Miller, and Max Vanzi. 2004, November. *Confined Animal Facilities in California.* California State Senate. http://www.sen.ca.gov/sor/reports/REPORTS_BY_SUBJ/ENVIRONMENT_NATURAL_RESOURCES/CAFFYI.pdf (accessed June 16, 2006).

Wu, F. 2006. Mycotoxin reduction in Bt corn: Potential economic, health and regulatory impacts. *Transgenic Research* 15: 277–289.

Xu, F. (ed.). 1992. *Encyclopedia of Chinese Customs (Zhongguo feng su ci dian).* Shanghai, China: Shanghai ci su chu ban she.

Ye, Xudong, Salim Al-Babili, Andreas Kloti, Jing Zhang, Paola Lucca, Peter Beyer, and Ingo Potrykus. 2000. Engineering the provitamin A (beta-carotene) biosynthetic pathway into (carotenoid-free) rice endosperm. *Science* 287: 303–305.

York Snow, Inc. 2001. Snowmax: York International. http://www.snowmax.com/products/snomax/ (accessed June 6, 2006).

CHAPTER 8

Balmford, Andrew, Aaron Bruner, Philip Cooper, Robert Costanza, Stephen Farber, Rhys E. Green, Martin Jenkins, Paul Jefferiss, Valma Jessamy, Joah Madden, Kat Munro, Norman Myers, Shahid Naeem, Jouni Paavola, Matthew Rayment, Sergio Rosendo, Joan Roughgarden, Kate Trumper, and R. Kerry Turner. 2002. Economic reasons for conserving wild nature. *Science* 297(5583): 950–953.

Brandes, Ray. 1970. *The Costanso Narrative of the Portola Expedition.* Newhall, CA: Hogarth Press.

Chivers, C. J. 2006. Putin urges plan to reverse slide in the birthrate. *New York Times.* May 11. http://www.nytimes.com (accessed June 5, 2006).

Conway, Gordon. 1997. *The Doubly Green Revolution.* Ithaca, NY: Cornell University Press.

Daily, Gretchen C. 2001. Ecological forecasts. *Nature* 411(May 17): 245.

Delgado, Christopher, Mark Rosegrant, Henning Steinfeld, Simeon Ehui, and Claude Courbois. 1999, October. Livestock to 2020: The next food revolution. *2020 Brief No. 61.* http://www.ifpri.org/2020/briefs/number61.htm (accessed April 17, 2006).

Donald, P. F., R. E. Green, and M. F. Heath. 2001. Agricultural intensification and the collapse of Europe's farmland bird populations. *Proceedings of the Royal Society of London. Series B: Biological Sciences* 286:25–29.

Durkin, Patrick R. 2003, March 1. Glyphosate—Human health and ecological risk assessment. Syracuse Environmental Research Association, Inc. Paper prepared for USDA Forest Service. http://www.fs.fed.us/r6/invasiveplant-eis/Risk-Assessments/04a03_glyphosate-final. pdf#xml=http://www.fs.fed.us/cgi-bin/texis/searchallsites/search.allsites/xml.txt?query= glyphosate+rain+water+november+17%2C+2000&db=allsites&id=424a94050 (accessed May 4, 2006).

Estrada, Alejandro, Rosamond Coates-Estrada, and Dennis A. Meritt, Jr. 1997. Anthropogenic landscape changes and avian diversity at Los Tuxtlas, Mexico. *Biodiversity and Conservation* 6(June 23): 19–43.

Estrada, Alejandro, Rosamond Coates-Estrada, Alberto Anzures Dadda, and Pierluigi Cammarano. 1998. Dung and carrion beetles in tropical rain forest fragments and agricultural habitats at Los Tuxtlas, Mexico. *Journal of Tropical Ecology* 14:577–593.

Extension Toxicology Network (EXTOXNET). 1996, June (revised). Diuron. Pesticide Information Profiles. A Pesticide Information Project of Cooperative Extension Offices of Cornell University, Oregon State University, the University of Idaho, and the University of California at Davis and the Institute for Environmental Toxicology, Michigan State University; USDA/Extension Service/National Agricultural Pesticide Impact Assessment Program. http://extoxnet.orst.edu/pips/diuron.htm (accessed May 8, 2006).

Fernandez-Cornejo, J., and W. D. McBride. 2000, April. *Genetically Engineered Crops for Pest Management in U.S. Agriculture: Farm-Level Effects.* Agricultural Economic Report 786. USDA Economic Research Service. http://www.ers.usda.gov/publications/aer786/aer786. pdf (accessed April 12, 2006).

Fernandez-Cornejo, Jorge, and Margriet Caswell. 2006. The First Decade of Genetically Engineered Crops in the United States. USDA Economic Research Service. Economic Information Bulletin, Number 11.

Giesy J. P., S. Dobson, and K. R. Solomon. 2000. Ecotoxicological risk assessment for Roundup herbicide. *Reviews of Environmental Contamination and Toxicology* 167: 35–120.

Global Amphibian Assessment (GAA). 2004, October. Summary of Key Findings. http://www.globalamphibians.org/summary.htm (accessed April 12, 2006).

Green, Rhys E., Stephen J. Cornell, Jörn P. W. Scharlemann, and Andrew Balmford. 2005. Farming and the fate of wild nature. *Science* 307(5709): 550–555.

Guthman, Julie. 2004. The trouble with 'organic lite' in California: A rejoinder to the 'conventionalisation' debate. *Sociologia Ruralis* 44(3): 301–316.

Hayes, Tyrone B., Atif Collins, Melissa Lee, Magdelena Mendoza, Nigel Noriega, A. Ali Stuart, and Aaron Vonk. 2002. Hermaphroditic, demasculinized frogs after exposure to the herbicide atrazine at low ecologically relevant doses. *Proceedings of the National Academy of Sciences. USA* 99(8): 5476–5480. http://www.pnas.org/cgi/content/

full/99/8/5476?maxtoshow=&HITS= 10&hits=10&RESULTFORMAT=&fulltext=atrazi ne&searchid=1&FIRSTINDEX=0&resourcetype=HWCIT (accessed May 8, 2006).

Huang, Jikun, Ruifa Hu, Scott Rozelle, and Carl Pray. 2005. Insect-resistant GM rice in farmers' fields: Assessing productivity and health effects in China. *Science* 296(April 29): 1694–1697.

Krebs, John R., Jeremy D. Wilson, Richard B. Bradbury, and Gavin M. Siriwardena. 1999. The second Silent Spring? *Nature* 400(August 12): 611–612.

Lappé, Frances Moore. 1971. *Diet for a Small Planet*. New York: Ballantine Books.

Maeder Paul, Andreas Fliessbach, David Dubois, Lucie Gunst, Padruot Fried, and Urs Niggli. 2002. Soil fertility and biodiversity in organic farming. *Science* 296: 1694.

Myers, Norman, and Jennifer Kent. 2003. New consumers: The influence of affluence on the environment. *Proceedings of the National Academy of Sciences USA* 100: 4963–4968.

Pain, D. J., and M. W. Pienkowski. 1997. *Farming and Birds in Europe: The Common Agricultural Policy and Its Implications for Bird Conservation*. London: Academic Press.

Pimentel, David S., and Peter H. Raven. 2000. Bt corn pollen impacts on nontarget Lepidoptera: Assessment of effects in nature. *Proceedings of the National Academy of Sciences USA* 97(15): 8198–8199.

Qaim, Matin, and David Zilberman. 2003. Yield effects of genetically modified crops in developing countries. *Science* 299(5608): 900–902.

Reganold J., J.Glover, P. Andrews, and H. Hinman. 2001. Sustainability of three apple production systems. *Nature* 410: 926.

Rosenzweig, Michael L. 2003a. *Win-Win Ecology: How the Earth's Species Can Survive in the Midst of Human Enterprise*. Oxford: Oxford University Press.

Rosenzweig, Michael L. 2003b. Reconciliation ecology and the future of species diversity. *Oryx* 37(2): 194–205.

Schaal, Barbara. 2007. Challenges for applications of transgenic plants from the perspective of population biology and environmental issues. Paper presented at the Arthur M. Sackler Colloquia of the National Academy of Sciences, From Functional Genomics of Model Organisms to Crop Plants for Global Health. Washington, D.C., April 3–5, 2006.

Snow, A. A., D. A. Andow, P. Gepts, E. M. Hallerman, A. Power, J. M. Tiedje, and L. L. Wolfenbarger. 2005. Genetically engineered organisms and the environment: current status and recommendations. *Ecological Applications* 15(2): 377–404.

Somerville, Chris, and John Briscoe. 2001. Genetic engineering and water. *Science* 292: 2217.

Steinfeld, H., P. Gerber, T. Wassenaar, V. Castel, M. Rosales, C. de Haan. 2006. *Livestock's Long Shadow; Environmental Issues and Options. Report of the Livestock, Environment and Development Initiative*. FAO United Nations, Rome: Publishing Management Service. http://www.virtualcentre.org/en/library/key_pub/longshad/A0701E00.htm

Strandberg, B., and M. B. Pederson. 2002. *Biodiversity in Glyphosate Tolerant Fodder Beet Fields*. Silkeborg: National Environmental Research Institute.

United Nations, Department of Economic and Social Affairs, Population Division. 2007. World Population Prospects: The 2006 Revision. http://www.un.org/esa/population/publications/wpp2006/wpp2006.htm

Waggoner, Paul E. 1995. How much land can ten billion people spare for nature? Does technology make a difference? *Technology in Society* 17(1): 17–34.

CHAPTER 9

Adam, David. 2003. Transgenic crop trials gene flow turns weeds into wimps. *Nature* 421: 462.

Adam, David. 2006. EU under attack over plan to ease organic labelling. *The Guardian*. January 6. http://www.guardian.co.uk/gmdebate/Story/0,,1680438,00.html (accessed April 18, 2006).

D. M. Arias and L. H. Rieseberg.1994. Gene flow between cultivated and wild sunflowers. *Theoretical and Applied Genetics* 89: 655–660.

Berthaud, Julien, and Paul Gepts. 2004. Assessment of effects on genetic diversity. In *Maize and Biodiversity: The Effects of Transgenic Maize in Mexico: Key Findings and Recommendations*. Article 13 Secretariat Report. North American Commission for Environmental Cooperation: Communications Department, CEC Secretariat. http://www.cec.org/maize/

Bollman, Marjorie Storm, George King, Peter K. Van de Water, Lidia S. Watrud, E. Henry Lee, Anne Fairbrother, Connie Burdick, and Jay R. Reichman. 2004. Evidence for landscape-level, pollen-mediated gene flow from genetically modified creeping bentgrass with CP4 EPSPS as a marker. *PNAS* 101: 14533–14538.

Byrne, P. F., and S. Fromherz. 2003. Can GM and non-GM crops co-exist? Setting a precedent in Boulder County, Colorado, U.S.A. *Journal of Food, Agriculture and Environment* 1(2): 258–261.

Cline, Harry. 2005. Growing, marketing herbicide-resistant alfalfa will be challenging, worthwhile. *Western Farm Press*, January 11.

Dahl, Roald. 1961. *James and the Giant Peach*. NY: Alfred A. Knopf, Inc.

Davis, S. D., V. H. Heywood, O. Herrera-MacBryde, J. Villa-Lobos, and A. Hamilton, (eds.). 1997. *Centres of Plant Diversity: A Guide and Strategy for Their Conservation, Volume 3. The Americas. Middle America and Carribean Islands: Sierra de Juarez, Oaxaca, Mexico*. Cambridge, UK: IUCN Publications Unit. http://www.nmnh.si.edu/botany/projects/cpd/ma/ma3.htm (accessed April 18, 2006).

Editor, *Nature*. 2002. Editorial Note. *Nature* 416(April 11): 600.

Ellstrand, Norman C. 2006. *Genetic Engineering and Pollen Flow*. Agricultural Biotechnology in California Series, Publication 8182. Davis, CA: The Regents of the University of California, Division of Agricultural and Natural Resources.

Garrett, Kelly Arthur. 2007. Transgenic pact signed by growers. *The Herald (Mexico)*, April 19. http://www.eluniversal.com.mx/miami/24289.html

Gil, Steven R., Mihai Pop, Robert T. DeBoy, Paul B. Eckburg, Peter J. Turnbaugh, Buck S. Samuel, Jeffrey I. Gordon, David A. Relman, Claire M. Fraser-Liggett, and Karen E. Nelson. 2006. Metagenomic analysis of the human distal gut microbiome. *Science* 312: 1355–1359.

GM Science Review Panel. 2003. *GM Science Review (First Report): An Open Review of the Science Relevant to GM Crops and Food Based on Interests and Concerns of the Public*. Department of Trade and Industry, London, UK. http://www.gmsciencedebate.org.uk/report.

Guyotat, Régis. 2005. La justice relaxe des faucheurs volontaires en invoquant le 'danger imminent' des OGM. *Le Monde,* December 11.

Halfhill, M. D., J. P. Sutherland, H. S. Moon, G. M. Poppy, S. I. Warwick, A. K. Weissinger, T. W. Rufty, P. .L. Raymer, and C. Neal Stewart, Jr. 2005. Growth, productivity, and competitiveness of introgressed weedy *Brassica rapa* hybrids selected for the presence of Bt *cry1Ac* and *gfp* transgenes. *Molecular Ecology* 14: 3177–3189.

Hawks, Bill. 2004, December 21. Letter to the National Association of State Departments of Agriculture. Washington, D.C.: Department of Agriculture, Office of the Secretary, Marketing and Regulatory Programs.

Hobhouse, Henry. 2005. *Seeds of Change: Six Plants That Transformed Mankind.* Emeryville: Shoemaker & Hoard.

National Research Council and Institute of Medicine of the National Academies. 2004. *Safety of Genetically Engineered Foods: Approaches to Assessing Unintended Health Effects.* Washington D.C.: The National Academy Press.

Ortiz-García, S., E. Ezcurra, B. Schoel, F. Acevedo, J. Soberón, and A. A. Snow. 2005. Absence of detectable transgenes in local landraces of maize in Oaxaca, Mexico (2003–2004). *Proceedings of the National Academy of Sciences* 102(35): 12338–12343.

Pimentel D., S. McNair, and J. Janecka. 2001. Economic and environmental threats of alien, plant, animal and microbial invasions. *Agriculture, Ecosystems and the Environment* 84: 1–20.

Pimentel D., L. Lach, R. Zuniga, D. Morrison. 2000. Environmental and economic costs of indigenous species in the United States. *Bioscience* 50: 53.

Pleasants J. M., R. L Hellmich., G. P. Dively, M. K. Sears, D. E. Stanley-Horn, H. R. Mattila, J. E. Foster, T. L. Clark, and G. D. Jones. 2001. Corn pollen deposition on milkweeds in and near cornfields. *Proceedings of the National Academy of Sciences USA* 98: 11919–11924.

Quist, D., and I. H. Chapela. 2001. Transgenic DNA introgressed into traditional maize landraces in Oaxaca, Mexico. *Nature* 414: 541–543.

Ross-Ibarra, Jeffrey, Peter L. Morrell, and Brandon S. Gaut. 2007. Plant domestication, a unique opportunity to identify the genetic basis of adaptation. *Proceedings of the National Academy of Sciences USA* 104: 8641–8648.

Schaal, Barbara. 2007. Challenges for applications of transgenic plants from the perspective of population biology and environmental issues. Paper presented at the Arthur M. Sackler Colloquia of the National Academy of Sciences, From Functional Genomics of Model Organisms to Crop Plants for Global Health. Washington, D.C., April 3–5, 2006.

Sears, M. K., and D. Stanley-Horn. 2000. Proceedings of International Symposium on the Biosafety of Genetically Modified Organisms. In C. Fairbairn, G. Scoles, and A. McHughen (eds.). *Proceedings of the 6th International Symposium on the Biosafety of Genetically Modified Organisms.* University of Saskatchewan, Canada: University Extension Press.

Strauss, S. 2003. Genomics, genetic engineering, and domestication of crops. *Science* 300: 61–62.

Trewavas, Anthony. 1999. Much food, many problems. *Nature* 402(November 18): 231–232.

Walz, Erika. 2003, May 16. 4th National Organic Farmers' Survey: Sustaining Organic Farms in a Changing Organic Marketplace. Organic Farming Research Foundation. http://www.ofrf.org/publications/survey/GMO.SurveyResults.PDF (accessed May 17, 2006).

Warwick, S. I. et al. (2003) Hybridization between transgenic *Brassica napus* L. and its wild relatives: *B. rapa* L., *Raphanus raphanistrum* L., *Sinapis arvensis* L., and *Erucastrum gallicum* (Willd.) O. E. Schulz. *Theoretical and Applied Genetics* 107: 528–539.

Watrud, L. S. et. al. 2004. Evidence for landscape-level, pollen-mediated gene flow from genetically modified creeping bentgrass with CP4 EPSPS as a marker. *Proceedings of the National Academy of Sciences USA* 101: 14533–14538.

Wilcove, D. S., D. Rothstein, J. Dubow, A. Phillips, and E. Losos. 1998. Quantifying threats to imperiled species in the United States. *Bioscience* 48(8): 607–615.

CHAPTER 10

California Deptartment of Pesticide Regulation. 2006. Summary of Results from the California Pesticide Illness Surveillance Program. HS-1865.

Federoff, Nina V., and Nancy Marie Brown. 2004. *Mendel in the Kitchen: A Scientist's View of Genetically Modified Food.* Washington D.C.: Joseph Henry Press.

GM Science Review Panel. 2003. *GM Science Review (First Report): An Open Review of the Science Relevant to GM Crops and Food Based on Interests and Concerns of the Public.* London: Department of Trade and Industry. http://www.gmsciencedebate.org.uk/report.

GRIN. 2006. http://www.ars.grin.gov/cgi-bin/npgs/html/pvplist.pl

Johnny's Selected Seeds. 2006 Catalog, p. 2.

National Corn Growers Association. 2004. NGCA Names Corn Yield Contest Winners. http://www.ncga.com/news/notd/2004/december/121504.html.

National Research Council and Institute of Medicine of the National Academies (NAS). 2004. *Safety of Genetically Engineered Foods: Approaches to Assessing Unintended Health Effects.* Washington D.C.: The National Academy Press.

Plant Variety Protection Office. http://www.ams.usda.gov/science/uvpo/PVPindex.htm

Tanksley, Steve, and Susan McCouch. 1997. Seed banks and molecular maps. Unlocking genetic potential from wild species. *Science* 277(5329): 1063–1066.

Vilmorin-Andrieux, M. M. 1885. *The Vegetable Garden.* London: John Murray. (Reprint 1976)

CHAPTER 11

AAAS Analysis of President's FY05 Budget Projections. 2004, May.

Atkinson, Richard C., Roger N. Beachy, Gordon Conway, France A. Cordova, Marye Anne Fox, Karen A. Holbrook, Daniel F. Klessig, Richard L. McCormick, Peter M. McPherson, Hunter R. Rawlings III, Rip Rapson, Larry N. Vanderhoef, John D. Wiley, and Charles E. Young. 2003. Intellectual property rights: Public sector collaboration for agricultural IP management. *Science* 301(July 11): 174–175.

Barahona, C., and Cromwell E. 2005. Starter pack and sustainable. In Sarah Levy (ed.). *Starter Packs: A Strategy to Fight Hunger in Developing Countries.* Oxford, UK: CAB International.

Bayh-Dole Act (University and Small Business Patent Procedures Act). 1980. 35 U.S.C. § 200–212. http://www.law.cornell.edu/uscode/html/uscode35/usc_sec_35_00000200----000-.html. Implemented by 37 C.F.R. 401, http://www.access.gpo.gov/nara/cfr/waisidx_02/37cfr401_02.html. Broothaerts, Wim, Heidi J. Mitchell, Brian Weir, Sarah Kaines, Leon M. A. Smith, Wei Yang, Jorge E. Mayer, Carolina Roa-Rodríguez, and Richard A. Jefferson. 2005. Gene transfer to plants by diverse species of bacteria. *Nature* (*Letters*) 433(February 10): 629–633. http://www.bios.net/daisy/bios/393/version/live/part/4/data (accessed June 9, 2006).

Brush, S. 1996. Whose knowledge, whose genes, whose rights? In S. Brush and D. Stabinsky (eds.). *Valuing Local Knowledge: Indigenous People and Intellectual Property Rights.* Washington D.C.: Island Press.

CAMBIA & IRRI. 2005, December 7. Open source biotechnology alliance for international agriculture: Mapping the patent maze to forge a shared research toolkit. Canberra and Los Baños. http://www.bios.net/daisy/bios/1374/version/default/part/AttachmentData/data/Open_Source_Biology_Alliance_for_International_Agriculture.pdf (accessed January 7, 2008).

Dennis, Sydney Carina. 2004. Biologists launch 'open-source movement.' *Nature* 431(September 30): 494.

Diamond v. Chakrabarty, 447 U.S. 303 (1980).

Henao, J., and Baanante C. A. March 2006. Agricultural production and soil nutrient mining in Africa. *Nature* 440: 728–729.

IRRI. 1990. Basic Facts about Rice. http://www.irri.org/about/faq.asp (accessed June 14, 2006).

Jacoby, C. D., and C. Weiss. 1997. *Stanford Environmental Law Journal* 16: 74.

Gladwell, M. 1995. Rights to life: Are scientists wrong to patent genes? *New Yorker*, November 13, pp. 120–124.

Khush, G. S., E. Bacalangco, and T. Ogawa. 1991. A new gene for resistance to bacterial blight from *O. longistaminata. Rice Genetics Newsletter* 7: 121–122.

Nakanishi, Nao. 2005. China Close to Production of 'Safe' Genetic Rice. Reuters, March 9.

Oldroyd, G. 2006. Nodules and hormones. *Science* 315: 52–53.

Public Intellectual Property Resource for Agriculture Newsletter. 2006. PIPRA's evaluation of the BiOS license. *Agriculture Newsletter* 5: 3.

Potrykus, Ingo. 2001. Golden rice and beyond. *Plant Physiology* 125(March): 1157–1161.

Richards, P. 1996. Culture and community values in the selection and maintenance of African rice. In: S. Brush and D. Stabinsky (eds.). *Valuing Local Knowledge: Indigenous People and Intellectual Property Rights.* Washington D.C.: Island Press.

Ronald, P. C., B. Albano, R. Tabien, L. Abenes, K. Wu, S. McCouch, and S. Tanksley. 1992. Genetic and physical analysis of the rice bacterial blight resistance locus, *Xa21. Molecular and General Genetics* 236: 113–120.

Sanchez, Pedro A. 2002. Soil fertility and hunger in Africa. *Science* 295(March 15): 2019–2020.

Song, W. Y., G. L. Wang, L. Chen, H. S. Kim, Li-Ya Pi, J. Gardner, B. Wang, T. Holsten, W. X. Zhai, L. H. Zhu, C. Fauquet, and P. Ronald. 1995. A receptor kinase-like protein encoded by the rice disease resistance gene *Xa21. Science* 270: 1804–1806.

Strauss, S. H. 2003. Genomics, genetic engineering, and domestication of crops. *Science* 300: 61–62.

ten Kate, Kerry, and Amanda Collis. 1997. Benefit-sharing case study: The GRRF. Submission to the Executive Secretary of the Convention on Biological Diversity by the Royal Botanic Gardens, Kew. http://indica.ucdavis.edu/publication/reference/csabs_ucdavis.pdf (accessed June 12, 2006).

Toenniessen, Gary H. 2006. Opportunities for and challenges to plant biotechnology adoption in developing countries. Paper delivered at the 15th annual meeting of the National Agricultural Biotechnology Council meeting, Science and Society at a Crossroad. Published by National Agricultural Biotechnology Council. http://www.rockfound.org/Library/

Opportunities_for_and_Challenges_to_Plant_Biotechnology_Adoption.pdf (accessed June 14, 2006).

CHAPTER 12

Alavanja, Michael C. R., Claudine Samanic, Mustafa Dosemeci, Jay Lubin, Robert Tarone, Charles F. Lynch, Charles Knott, Kent Thomas, Jane A Hoppin, Joseph Barker, Joseph Coble, Dale P. Sandler, and Aaron.Blair. 2003. Use of agricultural pesticides and prostate cancer risk in the agricultural health study cohort. *American Journal of Epidemiology* 157(9): 800–814.

Ammann, K. 2002.Thoughts about the future of agriculture: science and fiction in the risk assessment debate. In: International Food Policy Research Institute, *Sustainable Food Security for All by 2020: Proceedings of an International Conference, Bonn, Germany, September 4–6, 2001*. Washington, D.C.: IFPRI.

Berry, Wendell. 1987. *Home Economics: Fourteen Essays*. New York: North Point Press, Farrar, Straus and Giroux.

Campochiaro PA. 2006. Potential applications for RNAi to probe pathogenesis and develop new treatments for ocular disorders. *Gene Therapy* 13: 559–562.

Gonsalves, Dennis. 1998. Control of papaya ringspot virus in papaya: A case study. *Annual Review Phytopathology* 36: 415–37.

Haire, B. 2001. New Disease Threatens Georgia's Peach Industry. University of Georgia, College of Environmental and Agricultural Science. http://georgiafaces.caes.uga.edu/getstory.cfm?storyid=1241

Iadicicco R and Redding J. 2006. USDA seeks public comment on deregulation of genetically engineered plum. http://www.aphis.usda.gov/newsroom/content/2006/05/geplum.shtml

Kaplan, Kim. 2007. Plum-pox-resistant trees move forward. USDA Agricultural Research News Service. http://www.newsfood.com/Articolo/International/20070726-Plum-Pox-Resistant-trees-move-forward.asp

Kershen, Drew. 2006. Health and food safety: The benefits of Bt corn. *Food and Drug Law Journal* 61: 197.

Madison, Mike. 2006. *Blithe Tomato*. Berkeley, CA: Heyday Books.

McBride, J. 2006. One gene makes the difference—For plum pox resistance. USDA Agricultural Research Service. http://www.ars.usda.gov/is/AR/archive/sep01/gene0901.htm

Moses, Alan. 2006. Intelligent design: Playing with the building blocks of biology. *Berkley Science Review* 11: 34–40. http://sciencereview.berkeley.edu/articles/issue8/synthbio.pdf (accessed June 20, 2006).

Organic Consumers Association. 2006 USDA close to approving "frankenplums." http://www.organicconsumers.org/plum_alert.htm.

The Plant Disease Diagnostic Clinic. 2001, January. Plum Pox: *Plum Pox Virus* Factsheet. Cornell University. http://plantclinic.cornell.edu/FactSheets/plumpoxvirus/plumpox.htm (accessed June 20, 2006).

Pesticide Action Network North America (PANNA). 2005, August. "Critical use" exemptions—The methyl bromide loophole. *Global Pesticide Campaign* 15(2). http://www.panna.org/resources/gpc/gpc_200508.15.2.pdf (accessed May 19, 2005).

Pollan, Michael. 2007. *The Omnivore's Dilemma: A Natural History of Four Meals*. New York: Penguin Press.

Powell, D. A., K. Blaine, S. Morris and J. Wilson. 2003. Agronomic and consumer considerations for Bt and conventional sweet corn. *British Food Journal* 105(10): 700–713.

Shen, J., R. Samul, R. L. Silva, H. Akiyama, H. Liu, Y. Saishin, S. F. Hackett, S. Zinnen, K. Kossen, K. Fosnaugh, C. Vargeese, A. Gomez, K. Bouhana, R. Aitchison, P. Pavco, and P.A. Campochiaro. 2005. Suppression of ocular neovascularization with siRNA targeting VEGF receptor 1. *Gene Therapy* 13: 225–234.

Thomashow, M. F. 2001. So what's new in the field of plant cold acclimation? Lots! *Plant Physiology* 125: 89–93.

Toenniessen, Gary H. 2006. Opportunities for and challenges to plant biotechnology adoption in developing countries. Paper delivered at the 15th annual meeting of the National Agricultural Biotechnology Council meeting, Science and Society at a Crossroad. Published by National Agricultural Biotechnology Council. http://www.rockfound.org/Library/ Opportunities_for_and_Challenges_to_Plant_Biotechnology_Adoption.pdf (accessed June 14, 2006).

Index